Advances in Combustion of Gases, Liquid Fuels, Coal and Biomass

Advances in Combustion of Gases, Liquid Fuels, Coal and Biomass

Special Issue Editor

Peter B. Sunderland

MDPI • Basel • Beijing • Wuhan • Barcelona • Belgrade • Manchester • Tokyo • Cluj • Tianjin

Special Issue Editor
Peter B. Sunderland
University of Maryland
USA

Editorial Office
MDPI
St. Alban-Anlage 66
4052 Basel, Switzerland

This is a reprint of articles from the Special Issue published online in the open access journal *Energies* (ISSN 1996-1073) (available at: https://www.mdpi.com/journal/energies/special_issues/advances_combustion).

For citation purposes, cite each article independently as indicated on the article page online and as indicated below:

LastName, A.A.; LastName, B.B.; LastName, C.C. Article Title. *Journal Name* **Year**, *Article Number*, Page Range.

ISBN 978-3-03928-678-2 (Hbk)
ISBN 978-3-03928-679-9 (PDF)

Contents

About the Special Issue Editor

Peter B. Sunderland is Professor of Fire Protection Engineering at the University of Maryland. His research interests are in combustion and fire safety, including soot formation and oxidation, laminar diffusion flames, microgravity combustion, wildland fires, diagnostics development, refrigerant flammability, hydrogen flames, and vehicle fires. His degrees are from Cornell University (B.S.), the University of Massachusetts (M.S.), and the University of Michigan (Ph.D.). His awards include: NSF Career Award, two UMD teaching awards, Keystone Professor, the Isbrandt Automotive Safety Engineering Award, and Best-Paper Award from the Sixth International Seminar of Fire and Explosion Hazards. He is has co-authored of 55 journal articles and over 200 conference papers.

Preface to "Advances in Combustion of Gases, Liquid Fuels, Coal and Biomass"

This Special Issue of *Energies* on "Advances in Combustion of Gases, Liquid Fuels, Coal and Biomass" includes five manuscripts on combustion research related to energy production. Both fundamental and applied research is included. The papers contain state-of-the-art experiments, computations, and theory. Combustion provides an estimated 85% of the world's energy consumption. Advances in combustion research can benefit society in three main ways. Improving energy efficiency can reduce fuel consumption. Improving emissions can reduce climate change and adverse health effects. Improving fire and explosion safety can protect people, property, and the environment. The topical areas covered by this Special Issue are broad. It is hoped that this breadth will lead to a better understanding of combustion and improved diagnostic and numerical tools. This, in turn, may result in improved combustors, a cleaner environment, novel fuels, and improved safety and energy security.

Peter B. Sunderland
Special Issue Editor

energies

MDPI

Article

Experimental Investigation of Ash Deposit Behavior during Co-Combustion of Bituminous Coal with Wood Pellets and Empty Fruit Bunches

Tae-Yong Jeong [1], Lkhagvadorj Sh [1], Jong-Ho Kim [1], Byoung-Hwa Lee [2] and Chung-Hwan Jeon [1,3,*]

1 School of Mechanical Engineering, Pusan National University, 2, Busandaehak-ro 63 beon-gil, Geumjeong-gu, Busan 46241, Korea; nautilus0@pusan.ac.kr (T.-Y.J.); lkhagva_1166@pusan.ac.kr (L.S.); rlawhdgh2001@pusan.ac.kr (J.-H.K.)

2 Boiler R&D center, Doosan Heavy Industries and Construction, Volvoro 22, Seongsangu, Changwon 51711, Korea; byounghwa.lee@doosan.com

3 Pusan Clean Coal Center, Pusan National University, 2, Busandaehak-ro 63 beon-gil, Geumjeong-gu, Busan 46241, Korea

* Correspondence: chjeon@pusan.ac.kr; Tel.: +82-51-510-3051; Fax: +82-51-510-5236

Received: 17 April 2019; Accepted: 29 May 2019; Published: 31 May 2019

Abstract: In Korea, oil-palm empty fruit bunches (EFBs), which are byproducts of the crude palm-oil milling process, are among the most promising potential energy sources for power plants. However, the slagging and fouling characteristics of EFBs during combustion have not yet been fully studied. Accordingly, in this study, we investigated the fundamental ash behavior of EFBs in comparison to that of wood pellets (WPs) using a thermomechanical analyzer (TMA) and a drop-tube furnace (DTF). Ash melting and the deposition of ash particles were investigated with traditional prediction indices at several biomass blending ratios. The results demonstrated that, as the ratio of WPs to EFBs increases, the melting temperature decreases and the slagging propensity increases because of the increased biomass alkali content. Moreover, the penetration derived using the TMA shows a higher melting peak at which rapid melting occurs, and the melting temperature distribution is decreased with increased biomass blending. Conversely, the DTF results show different phenomena for ash deposition under the same blending conditions. Blend ratios approaching 10% WP and 15% EFB result in gradual decreases in ash deposition tendencies because of the lower ash contents of the co-combusted mass compared to that of the single coal ash. Further biomass addition increases ash deposition, which is attributable to ash agglomeration from the biomass. Thus, this study demonstrates that blending ratios of 10% WP and 15% EFB provide optimal conditions for co-combustion with the selected bituminous coal. In addition, it is shown that the slagging propensity of EFB is higher than that of WP owing to its ash content and simultaneous agglomeration.

Keywords: biomass; co-combustion; ash deposition; ash melting behavior; agglomeration

1. Introduction

The burning of fossil fuels is a major cause of climate change owing to the massive greenhouse gas (GHG) emissions it entails. Many countries signed the Paris Agreement at the UN Climate Change Conference in 2015 with the aim of reducing GHG emissions. Consequently, these countries are developing alternative and renewable energy sources for fossil fuel replacement [1]. Biomass continues to grow in importance worldwide as a renewable and CO_2-neutral energy source that may help to diversify renewable energy sources for energy production [2].

The South Korean government has announced its intention to reduce greenhouse gas emissions by 37% before 2030 [3]. To achieve this target, the South Korean government has recommended the use of new and renewable energies for power generation, such as biomass, by the relevant companies at the national level through the renewable portfolio standard (RPS). Furthermore, the government plans to increase the supply of new and renewable energy fuels from 3% in 2015 to 10% by 2023 [4].

Biomass co-combustion or co-firing with coal is a potential strategy to achieve GHG abatement and reduce toxic emissions. Many previous studies have demonstrated that biomass co-firing can reduce net CO_2 emission because of the carbon neutrality of biomass, as well as decreasing NO_x and SO_x production [5,6]. This technology is favorable for many power plants because the fuel can be easily used in existing powdered coal-powered boilers without entailing any environmental or economic concerns [7,8].

Although biomass co-combustion and co-firing have the advantages of simple equipment configuration, low cost, and enhanced combustion performance, biomass fuels usually contain high levels of alkali and alkaline metals, particularly Na and K, which increase slagging and fouling in biomass-fired boilers [9].

For example, Baxter [10] studied ash deposition and slagging/fouling during the combustion of coal and biomass, and postulated a mechanistic model and ash deposition characteristics for biomass combustion. According to his results, the ash deposition phenomenon is associated with the combustion conditions and the type of inorganic material used for fuel co-combustion. He found that ash deposition by biomass peaks during initial combustion and then gradually decreases.

Pronobis [11,12] investigated the effects of co-firing a medium-level fouling coal with three types of biomasses on surface fouling in the convection region of a furnace. He revealed that the properties of the biomass fuel affect both the operation variables and the efficiency of the boiler. Furthermore, he co-fired two types of bituminous coals with different slagging tendencies and four types of biomasses (straw, wood, dried sewage sludge, and bone meal) and examined the slagging effect in the furnace in terms of the correlation between the properties and fusibility of the produced ashes. The results showed that co-firing increases the risk of slagging at the fireside in furnaces.

Theis et al. [13] co-fired peat and two biomass types (bark and straw) using an entrained-flow reactor and compared the resulting ash deposition with that of single-fired peat. They reported that the ash deposition rate does not increase, even when the bark and straw fuels are co-fired with peat at levels of 30 wt% and 70 wt%, respectively.

Savolainen [14] researched the slagging behavior of boilers by measuring and monitoring the soot-blowing frequency and attemperation of water flow. She reported that slagging and fouling in the furnace were maintained at normal levels despite co-firing with biomass.

Abreu et al. [15] co-combusted bituminous coals with two types of biomasses (pine sawdust and olive stones) at 10–50 wt% (according to calorific value) and examined the ash deposition rate. Co-combustion sawdust with a low alkali content led to a lower ash deposition rate than that observed for single-firing of coal. In contrast, co-combustion with olive stones with high K contents resulted in a higher ash deposition rate than that for single coal. They posited that this difference is caused by the different adhesion tendencies on the deposition surface and attributed this effect to the ash compositions of the biomasses.

In addition to the above studies, numerous studies have been performed to understand and explain the phenomenon of ash deposition in biomass co-combustion. This includes the development of empirical indicators and several experimental methods for determining ash melting temperatures [16–18]. Many investigations using thermomechanical analyzers (TMAs), drop tube furnaces (DTF), pilot plants, and full-scale boiler trials have been performed [19–21].

These previous studies have demonstrated that slagging and fouling propensities vary, particularly in response to type of blended fuel used. Recently, the use of oil-palm empty fruit bunches (EFBs) as biomass has been attempted in power plants in Korea for economic reasons. Thus, optimal operation parameters for the application of EFBs in power plants must be derived. Oil palms are widely cultivated in tropical Asia, especially in Malaysia, Indonesia, and Thailand. Considering this abundant and

CO_2-neutral fuel resource, oil-palm EFBs, as byproducts of the crude palm-oil milling process, represent one of the most promising energy resources [22,23].

There is much less literature available on the ash deposition behavior of EFBs compared to that on wood pellets (WPs) [24,25]. Thus, the goal of this study was to perform a detailed investigation into ash deposition during the co-combustion of pulverized coal with EFBs and WPs in pulverized form.

A bituminous coal from Australia with a relatively low slagging propensity was chosen as a representative fuel that is commonly used in Korean thermal power plants. The alterations in the extents and mechanisms of coal slagging were investigated during co-combustion with the two biomass types (WP and EFB). The ash melting characteristics and deposition rates were analyzed using a TMA and a DTF. In addition, X-ray fluorescence (XRF) analysis was performed to examine the chemical compositions of the ashes and derive their empirical prediction indices.

2. Materials and Methods

2.1. Sample Preparation

For this study, one type of bituminous coal typically used at domestic coal-fired thermal power plants and the two biomass types were used as fuels. Trafigura coal, a bituminous coal imported from Australia, served as the coal sample, and the WP and EFB were procured from Canada and Malaysia, respectively. In this paper, bituminous coal is indicated as T coal.

The coal and biomass samples were pulverized using a ball mill and sieved to particle sizes of 75–90 μm and 400–600 μm. These particle sizes are commonly employed in coal-fired power plants in Korea. Before the experiments, proximate and ultimate analyses of the samples were conducted using a TGA-701 system (LECO Co., St. Joseph, MI, USA) and TruSpec element analyzer (LECO Co., St. Joseph, MI, USA) in accordance with the relevant ASTM standards. The procedures specified in the standards ASTM D5142 and ASTM D3176 for coal and in standards ASTM E870 for the two biomasses, respectively. The calorific values were measured using an AC600 (LECO Co., St. Joseph, MI, USA) calorimeter. Table 1 shows the sample analysis results.

Table 1. Properties of selected samples.

Contents	Proximate Analysis (%wt. [a], Dry Basis)			Ultimate Analysis (%wt., Dry Ash Free Basis)					Heating Value (LHV [b])
	VM [c]	Ash	FC [d]	Carbon	Hydrogen	Nitrogen	Oxygen [e]	Sulfur	MJ/kg
Trafigura (AUS)	29.72	17.11	53.17	80.00	5.56	1.27	12.41	0.76	24.92
WP (CAN)	80.40	0.29	19.31	47.29	6.39	–	46.30	0.02	18.65
EFB (MAS)	76.49	2.77	20.74	47.67	5.75	0.12	46.43	0.03	18.50

[a] AUS: Australia, CAN: Canada, MAS: Malaysia, %wt.: % by weight, [b] LHV: low heating value, [c] VM: volatile matter, [d] FC: fixed carbon, [e] Oxygen: by difference.

In addition, ashes were fabricated in accordance with ASTM E1755-01 to examine their chemical compositions depending on the experimental conditions and their melting characteristics through TMA experiments. Accordingly, 1 g of sample dried at 105 °C was placed in a ceramic crucible and heated in a temperature-controlled muffle furnace. The sample was heated in an O_2 atmosphere from room temperature to 250 °C at 10 °C/min, and the temperature was maintained for 30 min. Then, the sample was heated again to 575 °C (±25 °C) at 20 °C/min and the final temperature was maintained for 3 h to obtain ash samples, from which the carbon contents of the fuels were completely removed at a lower ash-producing temperature than that of ASTM D1857. This procedure was employed to avoid the volatilization of inorganic materials by the flames during the early combustion process [17]. The chemical composition of each fabricated ash sample was analyzed with a PW2400 XRF (Philips Co., Eindhoven, The Netherlands) at the Korea Basic Science Institute based on the ASTM D4326 method.

2.2. TMA Experiments

TMA experiments are typically conducted to investigate the melting behavior of ashes from solid fuels [18,26]. This method is known to improve the reproducibility of experiments by removing the effects of operators that may be present in conventional ash fusion tests such as ASTM D1857. This method provides two important types of data that can be used to identify the slagging propensity of a fuel. The first data type is the penetration tendency of the ram into the sample, which is expressed as a function of temperature. The second data type is the penetration rate according to temperature, which is expressed as "peak". The peak denotes the maximum penetration rate for a given temperature change. This peak indicates the temperature range within which the rapid melting of the ash sample occurs, and it is derived from the first derivative of the penetration trace [27].

Figure 1 shows a schematic of the TMA used in this study. The TMA apparatus is composed of a heating chamber and penetrating rod, as well as a ram, crucible, linear variable differential transformer (LVDT), and thermocouple. The experimental procedure used in this study was as follows:

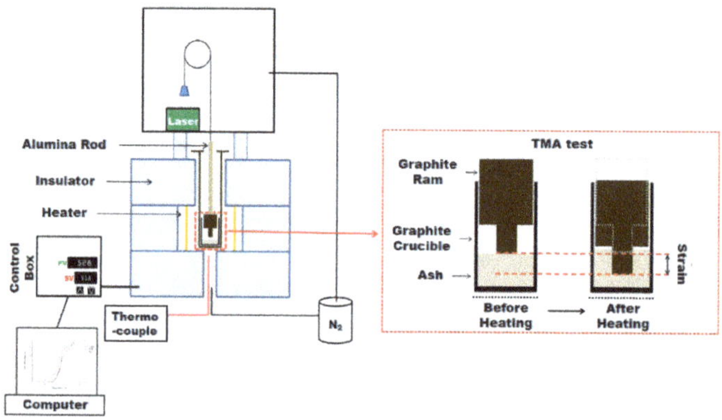

Figure 1. Schematic of the thermomechanical analyzer (TMA) apparatus.

Approximately 200 mg of an ash sample was placed in a crucible and the top of the sample was flattened with a jig by applying a constant pressure of 260 kPa. The sample was arranged in the prepared sample assembly by setting the pressure between the ram and the flattened sample surface at 30 kPa with a load of 60 g. The sample was then placed in a furnace. The sample was heated in a high-purity N_2 atmosphere. The furnace was first heated from room temperature to 600 °C at 50 °C/min, and then slowly heated at 5 °C/min from 600 °C to 1600 °C.

As shown in Figure 1, the ram sinks as the ash melts during the heating process, and the melted ash flows between the ram and the empty space in the bottom of the crucible when the ash is fully melted. The sinking displacement of the ram is measured with a laser displacement meter. This displacement indicates the height change of the ash sample from room temperature to a specific temperature and is converted to % shrinkage to evaluate the penetration of each sample according to temperature. The % shrinkage is used to identify the ash fusibility, which can be used as a predictor of ash deposition.

In addition, the peak at a specific temperature is determined from the first derivative of the % shrinkage. Generally, the peak distribution of a solid fuel shows more than two peaks, with each peak indicating rapid melting, i.e., melting acceleration intensity [28]. In this study, the peaks obtained from the shrinkage curves were differentiated at intervals of 10 min (or 50 °C) to prevent data scattering.

Gupta et al. [28] noted that temperatures of T25%, T50%, T75%, and T90% exhibit representative melting characteristics among the shrinkage values measured by the TMA in the melting of ash. As the label suggests, T25% refers to the temperature at which the ash sample reaches 25% shrinkage. In this case, the 25% (±15%) liquid phase appears by the softening and sintering of ash. Specifically, it can be regarded as the cause of the initial ash deposition growth because the ash particles are sticky. At T50%, the ash has shrunk by 50% and approximately 60% (+15%) melting occurs. At T75%, over 80% melting occurs, and this temperature is judged to denote complete melting. It is known that boilers must operate at temperatures lower than T75% to allow the discharge of fly ash without it melting. Finally, T90% is regarded as the final stage of melting (liquid phase > 90%). This temperature represents the slag flow (or fluid flow) characteristics of the ash.

2.3. DTF Experiments

To examine the ash deposition tendencies according to the type of coal and biomass co-combustion, a DTF apparatus (600 mm long with an internal diameter of 70 mm) was used [29]. Figure 2 shows schematics of the DTF apparatus and deposition probe. The DTF apparatus used in this study was designed to simulate operation variables such as heating rate and gas temperature applied to the pulverized fuel boiler in actual coal-fired thermal power plants and to mount an ash deposition probe to indicate the ash deposition characteristics of the solid fuels. The deposited ash particles were captured in an alumina tube (60 mm long with an outer diameter of 25 mm) with a hollow cylindrical shape mounted at the end of the probe and inserted perpendicular to the gas flow inside the DTF apparatus. The ash deposition probe was positioned 10 cm upstream from the back of the DTF apparatus. The particles captured by the probe had relatively low carbon contents (<1% unburned carbon) and were protected from heat by the coolant supplied externally to the water jacket inside the probe. The temperature inside the DTF was maintained at approximately 1300 °C, and the total gas flow rate, comprising a mixture of O_2 and N_2, was 5 L/min. Furthermore, the feeding rate of the sample into the DTF during deposition experiments was 0.2 g/min over 1 h, and this rate was chosen to compensate for supply errors made by the screw feeder mounted on top of the DTF. These errors are caused by two phenomena. The first is supply retention by the screw feeder at the beginning of the experiment due to the densely packed samples, and the second is rate reduction at the end of the experiment due to the loosely packed samples, leading to decreased self-loading of fuel particles. The rate of air transport was adjusted so that the excess O_2 content was maintained at 1.16 (*v/v*, dry), the typical furnace exit value for pulverized fuel boilers [29]. The mass of deposited ash particles was obtained by calculating the difference between the clean and fouled samples, and this value was used to derive the deposit growth rates.

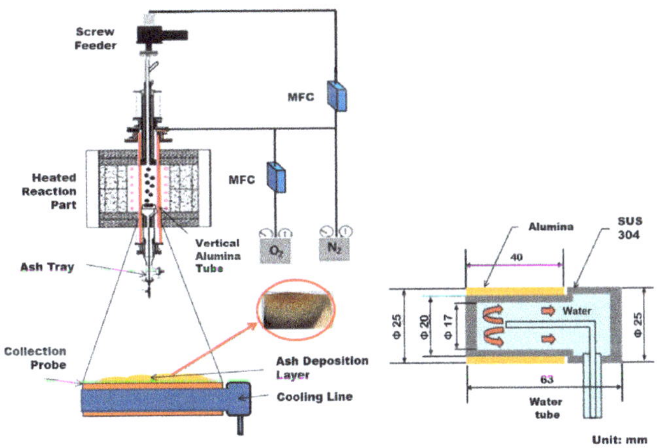

Figure 2. Schematic of the drop-tube furnace (DTF) apparatus and deposit probe (MFC: Mass flow controller; SUS: Steel use stainless).

In particular, to evaluate the ash deposition by biomass co-combustion at the same power output as that achieved with coal alone, the experiments were conducted based on the calorific value of

T coal, which is similar to that of the coal typically used in actual domestic power generation boilers. The relevant conditions are illustrated in Figure 3, and the blend conditions for coal and biomass were customized using the thermal fraction of the biomass, as seen in Equations (1)–(3).

$$T_{fB} = \frac{\dot{m}_B Q_B}{\dot{m}_B Q_B + \dot{m}_C Q_C}; \left(T_f = T_{fB} + T_{fC}\right), \quad (1)$$

$$M_{fB} = \frac{\dot{m}_B}{\dot{m}_B + \dot{m}_C}; \left(M_f = M_{fB} + M_{fC}\right), \text{ and} \quad (2)$$

$$A_{fB} = \frac{\dot{m}_B A_B}{\dot{m}_B A_B + \dot{m}_C A_C}; \left(A_f = A_{fB} + A_{fC}\right). \quad (3)$$

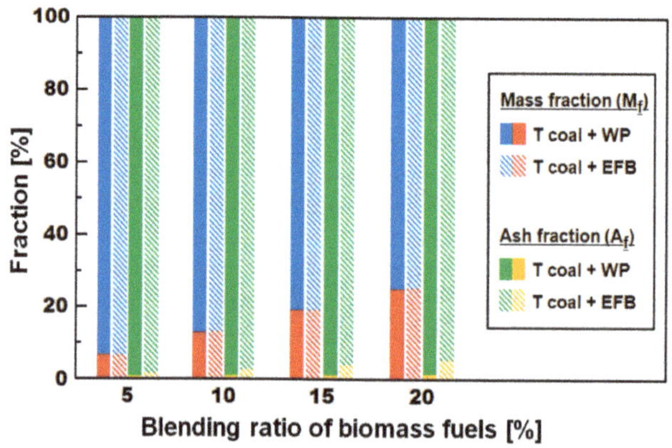

Figure 3. Blending conditions by thermal fractions of blends.

Here, T_f, M_f, and A_f denote the thermal, mass, and ash fractions by the proportions of the blends, respectively. Q, A, and $\dot{m}$ are the low heating value, ash content on dry basis, and mass flow (subscripts B and C denote biomass and coal), respectively.

Furthermore, the capture efficiency (CE) and energy-based growth rate (GRE) were derived to compare the ash deposition characteristics for the single and blended conditions from the deposition results obtained through five or more repeated experiments. The equations used for the calculations are as follows [30,31]:

$$CE = \frac{m_D}{m_A \times \left(\frac{A_C}{A_R}\right)} \times 100 \text{ and} \quad (4)$$

$$GRE = \frac{m_D}{LHV \times m_F}, \quad (5)$$

where m_D, m_A, and m_F denote the weight (g) of the deposit accumulated in the tube, total fly ash flowing towards the tube in the reactor, and blends fed during the test, respectively. A_C and A_R are the projected areas of the coupon and the cross-section of the DTF, respectively. These indices were derived from the weight of the collected particles acquired during the deposition experiment, and the measured weight was normalized to explain the difference in the fuel feeding rate.

3. Results and Discussion

3.1. Ashes from the Laboratory Experiments and Actual Combustion

The chemical composition of the ashes was analyzed to predict their deposition behaviors. In particular, to prevent the volatilization of the minerals generated from the initial biomass combustion process, separate analyses were conducted for the laboratory ash fabricated at a low temperature (575 °C) and the combustion ash obtained through the DTF experiment at a high temperature (1300 °C). For all the fuels fired, the char burnout in the combustion ash is always higher than 99% when the carbon content of the ash does not exceed 5%. The XRF results in Table 2 and Figure 4 show the chemical compositions of the laboratory/combustion ash and the propensity for slagging/fouling derived from several traditional indices. Table 3 presents the formulae and criteria pertaining to traditional prediction indices for ash deposition.

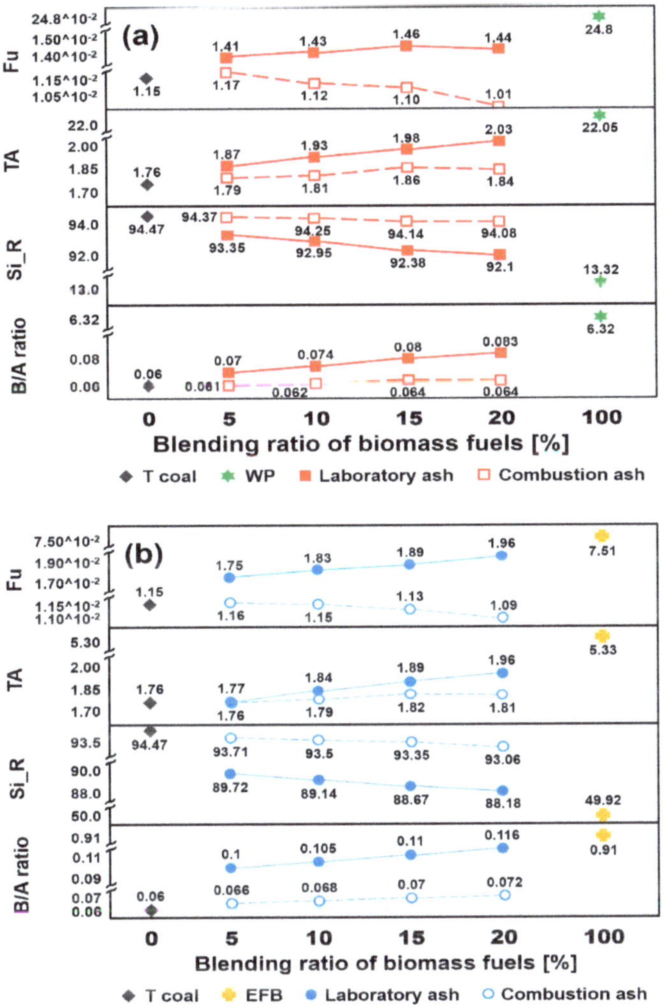

Figure 4. Trends of the traditional indices by X-ray fluorescence (XRF) results; (**a**) T coal + wood pellets (WP) and (**b**) T coal + empty fruit bunches (EFB) (B/A ratio, Si_R, TA, and Fu refer to base/acid ratio, silica percentage, total alkali, and fouling factor, respectively).

Table 2. Results of oxide analyses of laboratory and combustion ashes.

Components	Single Sample			T Coal + WP Blends				T Coal + EFB Blends			
	T Coal	WP	EFB	5%	10%	15%	20%	5%	10%	15%	20%
Laboratory ash											
SiO_2	66.27	8.75	39.30	65.72	65.43	65.13	64.92	63.39	63.11	62.79	62.44
Al_2O_3	26.69	3.54	9.22	26.31	26.25	26.07	25.94	26.09	25.81	25.62	25.43
TiO_2	1.24	0.25	0.45	1.21	1.19	1.18	1.22	1.26	1.25	1.29	1.31
Fe_2O_3	3.10	4.68	31.82	3.10	3.22	3.34	3.41	6.44	6.72	6.93	7.16
CaO	0.42	43.92	4.52	1.18	1.33	1.59	1.67	0.43	0.51	0.58	0.64
MgO	0.36	8.62	3.09	0.40	0.41	0.44	0.49	0.39	0.46	0.51	0.57
Na_2O	0.08	0	0	0.07	0.06	0.05	0.05	0.05	0.04	0.03	0.02
K_2O	1.68	22.05	5.33	1.80	1.87	1.93	1.98	1.72	1.80	1.86	1.94
P_2O_5	0.16	4.99	3.01	0.16	0.16	0.17	0.18	0.17	0.19	0.22	0.25
SO_3	0	3.20	3.26	0.05	0.08	0.10	0.14	0.06	0.11	0.17	0.24
Combustion ash											
SiO_2				66.16	66.10	66.01	65.96	65.88	65.71	65.57	65.39
Al_2O_3				26.69	26.64	26.60	26.62	26.51	26.43	26.41	26.33
TiO_2				1.22	1.23	1.23	1.22	1.23	1.24	1.22	1.22
Fe_2O_3				3.10	3.11	3.11	3.12	3.67	3.78	3.84	4.01
CaO		n.a. [a]		0.48	0.55	0.62	0.65	0.40	0.43	0.44	0.46
MgO				0.37	0.37	0.38	0.38	0.35	0.36	0.39	0.41
Na_2O				0.07	0.06	0.06	0.05	0.07	0.06	0.05	0.03
K_2O				1.72	1.75	1.80	1.79	1.69	1.73	1.77	1.78
P_2O_5				0.18	0.18	0.19	0.19	0.17	0.21	0.24	0.26
SO_3				0.01	0.01	0.01	0.02	0.03	0.05	0.07	0.11

[a] n.a.: Not analyzed.

Table 3. Summary of traditional ash deposition indices and associated criteria.

Indices	Formula	Criteria			
		Low	Medium	High	Severe
B/A ratio [32]	$(Fe_2O_3 + CaO + MgO + Na_2O + K_2O)/(SiO_2 + Al_2O_3 + TiO_2)$	<0.5	0.5–0.7	0.7–1.0	>1.0
Si_R [32]	$(SiO_2 \times 100)/(SiO_2 + Fe_2O_3 + CaO + MgO)$	>50	30–50	5.0–30	<5.0
TA [32]	$Na_2O + K_2O$	<2.0	2.0–3.0	3.0–4.0	>4.0
Fu [20]	$0.01 \times A_p$ [a] $\times$ B/A ratio $\times (Na_2O + 0.659 \times K_2O)$	<0.3	0.3–0.45	0.45–0.60	>0.6

[a] A_p: mass percentage of ash in fuel (or blends).

For the laboratory ash, the basic oxides are enriched as the blending ratio is increased, as shown in Table 2. This result is attributed to the unreacted mineral components remaining as-is (un-volatilized) at the low temperature. The effects of basic oxides from both biomass ashes are more evident using the traditional predictive indices and consequently affect the base/acid (B/A) ratio, total alkali (TA), and fouling factor (Fu) of the laboratory ash. As shown in Figure 4, the trends of indices derived from the XRF analysis using laboratory ash show that the slagging/fouling propensity continuously increases as the blending ratio of the biomass is increased.

Conversely, for the combustion ash obtained from the DTF experiments, acid oxides of the deposited ash samples are increased slightly compared to those of the laboratory ash. Particularly, the trends in B/A ratio and Fu identified in Figure 4 have very similar values within the criteria of Table 3, regardless of ash type. This is thought to correspond to the characteristics of the coal ash imparted by the minute fraction of biomass ash in the blending sample, although there is also an effect from the basic oxide of the biomass ash. The combustion ash is a good representation of a fly ash generated by a boiler. The combustion ash of the T coal blended with biomass mainly contains silica and alumina oxides (more than 90 wt%). Nevertheless, the chemical compositions of the laboratory ash and combustion ash vary only slightly with increasing biomass blending ratio, and neither ash type is significantly different from the T coal ash under the given blend conditions. This finding is in agreement with that of Kupka et al. [21], who observed that, excepting sulfur oxides, fly ash and crucible ash compositions, expressed as oxides, are generally consistent. Thus, Pronobis [12] proposed that the chemical composition of biomass during the co-combustion process is not clearly distinguishable from that of coal when the blending ratio for the thermal fraction of the biomass does not exceed 20%. In other words, the result implies that the chemical composition of biomass ash does not influence that of the blend when the amount of biomass supplied is sufficiently small compared with that of the coal in terms of thermal input.

Tables 2 and 3 show that the deposition tendencies for both types of ashes meet the low-level criterion within each prediction index range. However, as the addition of biomass increases, the evaluations of the traditional prediction indices gradually increase, especially the TA index for the WP 20% blend. This implies that the blending ratio of coal and biomass whose basic oxides content reaches 90%, such as in the case of WP, should be carefully selected. The XRF results of this study show a difference between the chemical compositions of the ash types, which were generated at different temperatures according to the biomass co-combustion conditions, while the traditional prediction index results do not show a clear difference. Thus, a distinct deposition tendency cannot be confirmed. It is found, however, that as the biomass blending ratio increases, the deposition tendency is slightly increased regardless of ash type because the basic oxides are derived from the biomass ash.

As remarked by Kazagic and Smajevic [20], ash deposition evaluation based on laboratory ash analysis, including the calculation of traditional indices, shows poor reliability for biomasses, and laboratory results cannot be accepted as sufficiently reliable for evaluating the slagging and fouling propensities of biomasses. Therefore, this study performed further experiments using improved methods such as the TMA and DTF, as explained in the following sections.

3.2. Ash Melting Characteristics from TMA Experiments

To characterize the melting behavior of the laboratory ash, TMA experiments were performed for single-coal and blended samples. Figure 5 shows the shrinkage curves as a function of temperature for each raw sample.

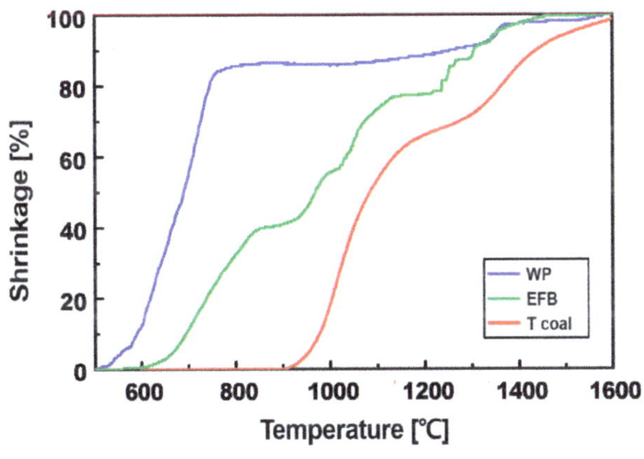

Figure 5. Shrinkage curves as a function of temperature for each raw sample.

The shrinkage curves shown in Figure 5 provide a good representation of the melting characteristics of each raw sample according to temperature. T coal is not completely melted at 1600 °C, which is the upper limit of the temperature in the TMA experiments performed. This is because T coal is a typical bituminous coal with acid oxide contents exceeding 90%, as seen from the XRF results [33].

Conversely, as demonstrated by the melting traces for both raw biomass samples, their melting reactions are completed at lower temperatures compared to that of T coal. For WP, the basic oxides content exceeds 85%, and T90% appears at a temperature lower than 1150 °C. For EFB, T90% appears at approximately 1300 °C, i.e., at a higher temperature than that of WP. This result is attributed to the ash components of WP, which have higher alkali contents than EFB, as shown in Table 2 and Figure 4.

When ash samples of blends are heated according to the thermal fraction, the melting trace of each blend follows that of raw T coal, which begins to melt at approximately 900 °C, as shown in Figure 6.

The melting temperatures for blended biomass are slightly lower or similar to that of raw T coal when the shrinkage is lower than T50% at all blending ratios. Conversely, when the shrinkage is higher than T50%, the melting temperatures are lower than that of raw T coal under all experimental conditions. It appears that the shrinkage is increased because the melting of raw biomasses is almost complete at temperatures lower than approximately 1200 °C, at which they are converted to the liquid phase. Moreover, this tendency is conspicuous in the results for EFB and when comparing the T75% value for the blend conditions; the decreased temperature for EFB blending is approximately 100 °C greater than that of the WP blends because EFB has higher ash contents despite its lower alkali content.

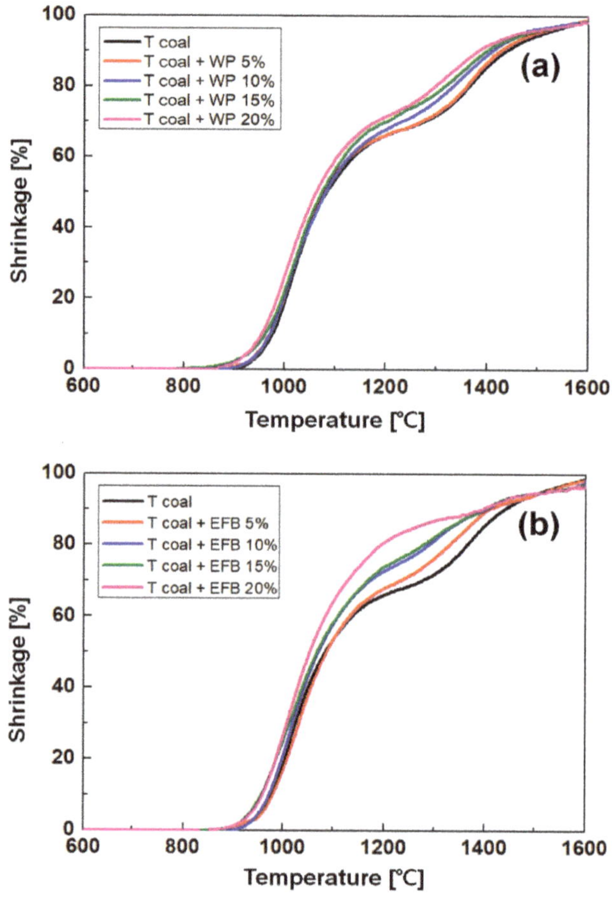

Figure 6. Shrinkage curves as a function of temperature for different blending ratios. (**a**) T coal + WP; (**b**) T coal + EFB.

Figure 7 illustrates the derivative trace of the results seen in Figure 6a,b. These graphs show the melting peak at which rapid melting occurs identified from the first derivative of the melting results derived from the shrinkage of biomass blending and the change in the peak temperature. The melting peaks of the biomass blends revealed in Figure 7a,b occur before approximately 1050 °C, which corresponds to a value between T25% and T50% of T coal.

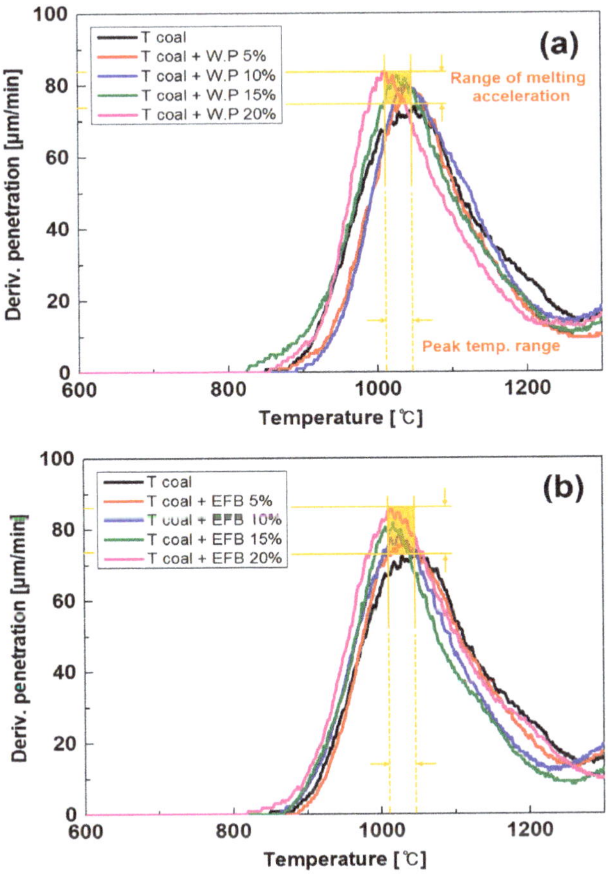

Figure 7. Derivative traces of the blend conditions derived from the TMA results; (**a**) T coal + WP; (**b**) T coal + EFB.

As shown in Figure 7, the melting peaks for the blends of both biomasses shift towards lower temperatures as the blending ratio increases based on the peak of T coal (at approximately 1050 °C). These peak temperatures are distributed between T25% and T50%, which is the shrinkage characteristic temperature of T coal. The melting is accelerated for the biomass as the peak height is gradually increased. Furthermore, in Figure 7a,b, the yellow peak distribution boxes indicate the melting intensity under a given blending ratio. The range of peak temperatures for both blends are not significantly different. However, the peak heights for EFB blends are higher than those of WP blends in a similar range of peak temperatures. Herein, the peak temperature is associated with the chemical composition of a sample and the height of the peak indicates a specific melting intensity. Since there is no difference in the chemical compositions of the ash of biomass blends and laboratory ash by XRF analysis, the melting intensity is affected by ash content according to blending conditions. As a result, the melting traces for biomass blending obtained with the TMA are confirmed. The melting temperature steadily decreases as the blending ratio of biomass increases, and this phenomenon is not reversed beyond a certain blending ratio. Therefore, this indicates that, as the biomass content in the blend increases, the melting temperature decreases and the slagging/fouling tendency can be expected to increase gradually.

3.3. Ash Deposition Characteristics from DTF Experiments

The DTF was used to examine the ash deposition of coal and biomass according to the experimental conditions. The deposited ash particles collected on the deposition probe during 1 h of experimentation and the deposition tendencies derived by the CE and GRE were used as the existing coal deposition indices. Figure 8 shows that co-combustion with WP and EFB generate relatively lower ash depositions than that of T coal alone. The single-coal deposition is evenly distributed over the entire area of the coupon and the deposited particles are stacked in a convex shape. In the case of biomass co-combustion, however, there is less deposition than in the case of single coal, and a narrow deposition distribution, with relatively low deposition on the side of the coupon, is observed along with a flatter deposit. Although these deposition images appear to show that the deposition is decreased compared to that of T coal as the blending ratio of the biomass is increased, it is difficult to confirm distinct trends.

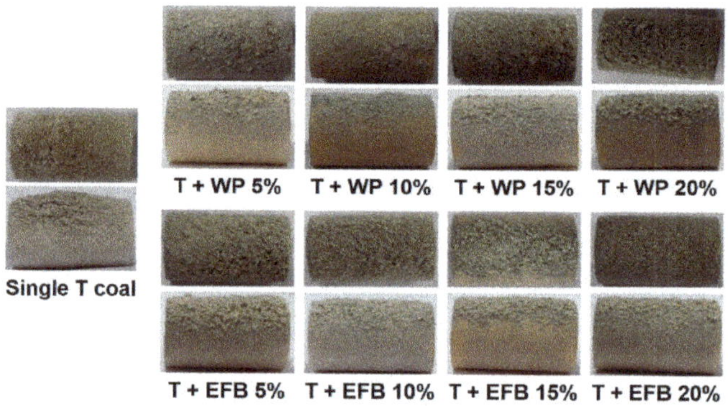

Figure 8. Images of ash collected on the deposit probe for different blend conditions and single combustion.

Figure 9 shows all the CE and GRE values obtained by co-combustion the biomass for all the blend conditions mentioned in Figure 3. The results show that, as the blending ratio of biomass (WP and EFB) increases, the deposition rate decreases for the 10% and 15% blends, but it begins to increase again with additional blending. In general, it is known that the ash in biomass contains significant amounts of alkali matter, which leads to a lower melting temperature, as shown in the TMA results. Most previous studies [34–36] reported that the addition of biomass increases ash deposition and attribute this result to the lower melting temperature and higher stickiness of biomass. Namkung et al. [37] also remarked that sticky particles tend to adhere to each other because of the partial melting characteristics of the particle surface and that increased biomass blending enhances ash deposition and agglomeration behavior. However, even though the blending ratio is increased, the deposition is initially decreased to a certain extent. This is why the ash contents of the biomasses (0.29% for WP and 2.77% for EFB) are much lower than that of T coal ash (17.11%). This finding is in agreement with those of previous studies. For example, Abreu et al. [15] reported that the co-combustion of coal with lower ash biomass does not pose operational problems related to the occurrence of slagging and fouling because the deposition rate decreases with increases in the biomass thermal input in the blend. In addition, similar effects have been reported by Kupka et al., who indicated that the ash deposition rate is influenced by ash input [21]. Therefore, these results point to the existence of additive and non-additive phenomena caused by lower ash contents and higher agglomeration effects during the co-combustion of coal with the biomasses.

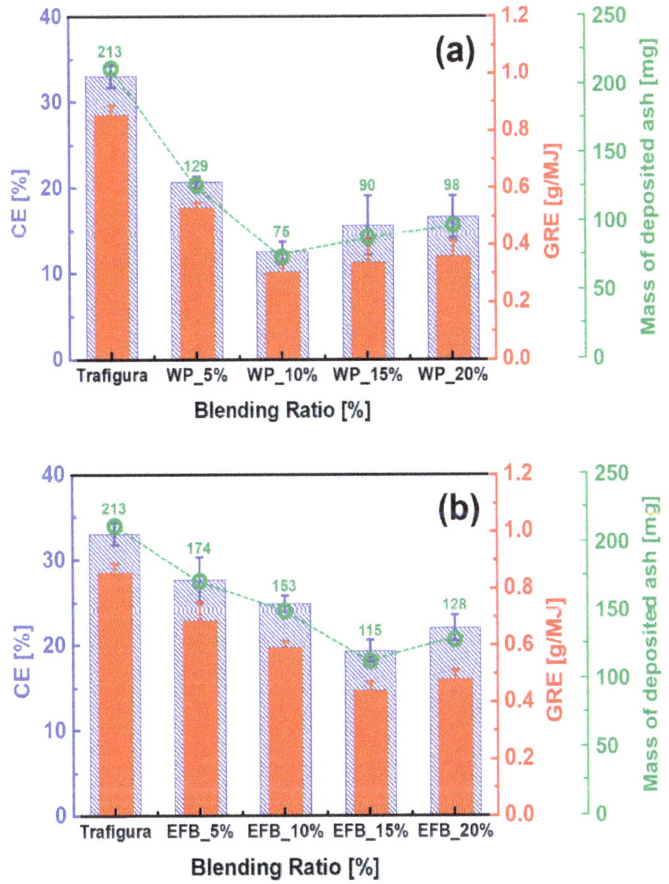

Figure 9. Values of capture efficiency (CE) and energy-based growth rate (GRE) for the blend conditions in the DTF; (**a**) T coal + WP; (**b**) T coal + EFB.

Figure 10 shows the normalized GRE values for the blend conditions in the DTF. Herein, GRE_{max} represents the ratio of the GRE values of other samples to the maximum GRE value. The GRE_{max} range of the WP blends is approximately 0.35–0.60, and that of the EFB blends is approximately 0.50–0.80. While the deposition of WP blends is less than half the deposition by the single combustion of T coal for all blending ratios except 5%, the deposition of EFB blends is higher than half the T coal deposition under all the co-combustion conditions. These results show that the deposition tendency with the co-combustion of EFB is higher than that of WP, unlike the predicted results using the chemical composition of the biomass revealed by XRF. In other words, the ash amounts of EFB, according to the feeding conditions, are approximately 10 times higher than those of WP, which demonstrates the influence of the deposition tendency. In addition, the reversal tendency of the EFB blend, i.e., the increase in deposition after the decrease, is changed at a higher blending ratio. This indicates that the deposition is more influenced by the feeding rate of the EFB than by its chemical composition when the amount of the biomass supplied is sufficiently small.

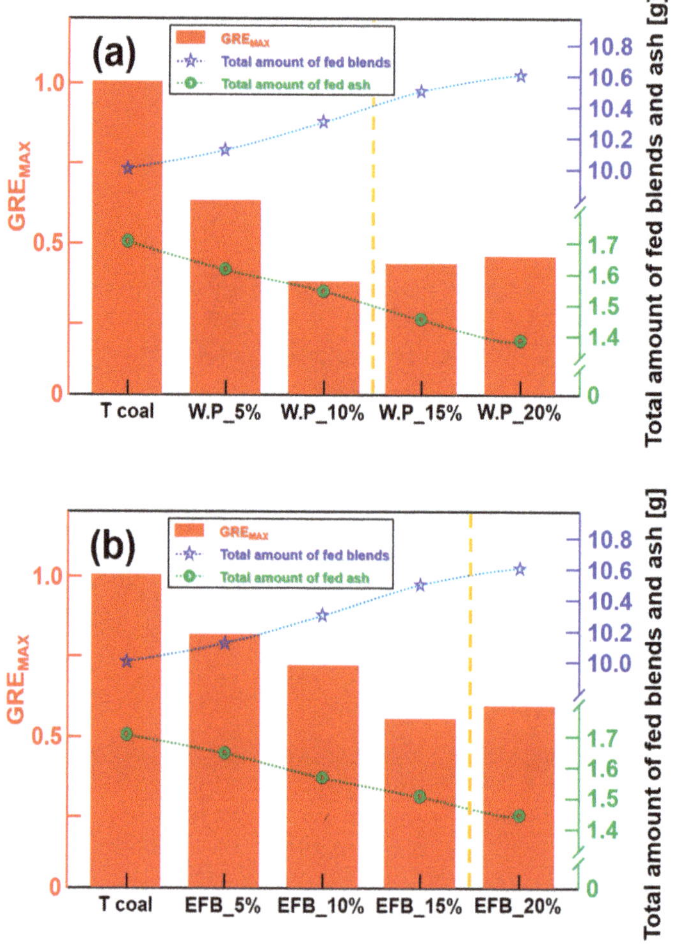

Figure 10. Normalized GRE values for blend conditions in the DTF; (**a**) T coal + WP; (**b**) T coal + EFB.

4. Conclusions

This study aimed to fundamentally investigate the ash deposition behavior during co-combustion of coal with two kinds of biomasses, WP and EFB, using TMA and DTF. The conclusions are as follows.

1. The laboratory and combustion ashes were analyzed using XRF and the slagging indices were derived using the chemical compositions. The ash compositions as per the results of XRF showed that the basic oxides were enriched as the blending ratio increased. Thus, the results of the XRF analysis showed that the deposition tendency continuously increases with an increasing biomass blending ratio, but the traditional indices for ash slagging propensity were insufficient for explaining the results because of their predictive limitations.
2. Ash melting characteristics were examined for the laboratory ash using a TMA. The % shrinkage traces of the raw fuel showed the individual melting characteristics of each fuel, and their % shrinkage traces for blends at T25% appeared to follow the melting characteristics of T coal. Conversely, the melting point clearly decreased after T50% as the blending ratio of the biomass increased. This indicates that the shrinkage accelerates when the melting of the biomass is almost complete and the biomass turns into a liquid phase. The TMA results revealed a decrease in

melting temperature with increasing WP and EFB blending, and the slagging/fouling tendency was expected to increase gradually.

3. The ash deposition characteristics revealed using the DTF showed a lower deposition tendency than that of single-coal combustion up to a blending ratio of 10% for WP and 15% for EFB. However, with further blending, a reversal was observed wherein the deposition tendency gradually increased. These results indicated the occurrence of additive and non-additive phenomena caused by the lower ash contents and higher agglomeration effects during the co-combustion of coal with the biomass.
4. These results indicate that an optimum blending ratio exists for the co-combustion of biomass with coal, and considering the ash contents as well as the agglomeration caused by the biomass ash indicated that a blending ratio of 10% for WP and 15% for EFB are the optimal conditions. The slagging tendency of EFB was worse than that of WP because of the ash contents of EFB.
5. WP have been researched extensively in Korea, but plant application of EFB can now be attempted. This study was performed in order to provide fundamental analysis and experiments using laboratory-scale apparatus, but further studies such as pilot and plant tests should be performed before applying EFB-derived fuels in power plants.

Author Contributions: Conceptualization, T.-Y.J. and C.-H.J.; data curation, T.-Y.J. and C.-H.J.; investigation, T.-Y.J., L.S., and J.-H.K.; methodology, T.-Y.J. and B.-H.L.; project administration, C.-H.J.; supervision, C.-H.J.; validation, T.-Y.J., L.S., J.-H.K., and B.-H.L.; visualization, T.-Y.J. and L.S.; writing—original draft, T.-Y.J.; writing—review and editing, T.-Y.J., B.-H.L., and C.-H.J.

Acknowledgments: This work was conducted under the framework of the research and development program of the Korea Institute of Energy Research (B9-2445).

Conflicts of Interest: The authors declare no conflict of interest.

References

1. Huang, C.W.; Li, Y.H.; Xiao, K.L.; Lasek, J. Cofiring characteristics of coal blended with torrefied Miscanthus biochar optimized with three Taguchi indexes. *Energy* **2019**, *172*, 566–579. [CrossRef]
2. Fuller, A.; Omidiji, Y.; Viefhaus, T.; Maier, J.; Scheffknecht, G. The impact of an additive on fly ash formation/transformation from wood dust combustion in a lab-scale pulverized fuel reactor. *Renew. Energy* **2019**, *136*, 732–745. [CrossRef]
3. Jeong, Y.-S. Assessment of alternative scenarios for CO_2 reduction potential in the residential building sector. *Sustainability* **2017**, *9*, 394. [CrossRef]
4. Park, H.-G.; Kim, C.-H. Do shifts in renewable energy operation policy affect efficiency: Korea's shift from FIT to RPS and its results. *Sustainability* **2018**, *10*, 1723. [CrossRef]
5. Lasek, J.A.; Kazalski, K. Sulfur self-retention during co-combustion of fossil fuels with biomass. *Energy Fuels* **2014**, *28*, 2780–2785. [CrossRef]
6. Liu, H.; Gibbs, B.M.; Hampartsoumian. The significance of rank on coal reburning for the reduction of NO in drop tube furnace. In Proceedings of the 8th International Symposium on Transport Phenomena in Combustion, San Francisco, CA, USA, 16–20 July 1995; pp. 329–340.
7. Lazaroiu, G.; Frentiu, T.; Mihaescu, L.; Mihaltan, A.; Ponta, M.; Frentiu, M.; Cordos, E. The synergistic effect in coal/biomass blend briquettes combustion on elements behavior in bottom ash using ICP-OES. *J. Optoelectron. Adv. M.* **2009**, *11*, 713–721.
8. Ali, U.; Akram, M.; Font-Palma, C.; Ingham, D.B.; Pourkashanian, M. Part-load performance of direct-firing and co-firing of coal and biomass in a power generation system integrated with a CO_2 capture and compression system. *Fuel* **2017**, *210*, 873–884. [CrossRef]
9. Shao, Y.; Wang, J.; Preto, F.; Zhu, J.; Xu, C. Ash deposition in biomass combustion or co-firing for power/heat generation. *Energies* **2012**, *5*, 5171–5189. [CrossRef]
10. Baxter, L.L. Ash deposition during biomass and coal combustion: A mechanistic approach. *Biomass Bioenerg.* **1993**, *4*, 85–102. [CrossRef]

11. Pronobis, M. The influence of biomass co-combustion on boiler fouling and efficiency. *Fuel* **2006**, *85*, 474–480. [CrossRef]
12. Pronobis, M. Evaluation of the influence of biomass co-combustion on boiler furnace slagging by means of fusibility correlations. *Bioresour. Bioenerg.* **2005**, *28*, 375–383. [CrossRef]
13. Theis, M.; Skrifvars, B.J.; Hupa, M.; Tran, H. Fouling tendency of ash resulting from burning mixtures of biofuels. Part 1: Deposition rates. *Fuel* **2006**, *85*, 1125–1130. [CrossRef]
14. Savolainen, K. Co-firing of biomass in coal-fired utility boilers. *Appl. Energy* **2003**, *74*, 369–381. [CrossRef]
15. Abreu, P.; Casaca, C.; Costa, M. Ash deposition during the co-firing of bituminous coal with pine sawdust and olive stones in a laboratory furnace. *Fuel* **2010**, *89*, 4040–4048. [CrossRef]
16. Sahu, S.G.; Chakraborty, N.; Sarkar, P. Coal-biomass co-combustion: An overview. *Renew. Sustain. Energy Rev.* **2014**, *39*, 575–586. [CrossRef]
17. Chen, X.; Tang, J.; Tian, X.; Wang, L. Influence of biomass addition on Jincheng coal ash fusion temperatures. *Fuel* **2015**, *160*, 614–620. [CrossRef]
18. Rushdi, A.; Sharma, A.; Gupta, R. An experimental study of the effect of coal blending on ash deposition. *Fuel* **2004**, *83*, 495–506. [CrossRef]
19. Febrero, L.; Granada, E.; Regueiro, A.; Miguez, J.L. Influence of combustion parameters on fouling composition after wood pellet burning in a lab-scale low-power boiler. *Energies* **2015**, *8*, 9794–9816. [CrossRef]
20. Kazagic, A.; Smajevic, I. Experimental investigation of ash behavior and emissions during combustion of Bosnian coal and biomass. *Energy* **2007**, *32*, 2009–2016. [CrossRef]
21. Kupka, T.; Mancini, M.; Irmer, M.; Webber, R. Investigation of ash deposit formation during co-firing of coal with sewage sludge, saw-dust and refuse derived fuel. *Fuel* **2008**, *87*, 2824–2837. [CrossRef]
22. Pimenidou, P.; Dupont, V. Characterisation of palm empty fruit bunch (PEFB) and pinewood bio-oils and kinetics of their thermal degradation. *Bioresour. Technol.* **2012**, *109*, 198–205. [CrossRef] [PubMed]
23. Madhiyanon, T.; Sathitruangsak, P.; Sungworagarn, S.; Pipatmanomai, S.; Tia, S. A pilot-scale investigation of ash and deposition formation during oil-palm empty-fruit-bunch (EFB) combustion. *Fuel Process. Technol.* **2012**, *96*, 250–264.
24. Guo, F.; Zhong, Z. Co-combustion of anthracite coal and wood pellets: Thermodynamic analysis, combustion efficiency, pollutant emissions and ash slagging. *Environ. Pollut.* **2018**, *239*, 21–29. [CrossRef]
25. Zeng, T.; Pollex, A.; Weller, N.; Lenz, V.; Nelles, M. Blended biomass pellets as fuel for small scale combustion appliances: Effect of blending on slag formation in the bottom ash and pre-evaluation options. *Fuel* **2018**, *212*, 108–116. [CrossRef]
26. Kim, G.-M.; Lee, D.-G.; Jeon, C.-H. Fundamental characteristics and kinetic analysis of lignocellulosic woody and herbaceous biomass fuels. *Energies* **2019**, *12*, 1008. [CrossRef]
27. Liu, Y.; Gupta, R.; Elliot, L.; Wall, T.; Fujimori, T. Thermomechanical analysis of laboratory ash, combustion ash and deposits from coal combustion. *Fuel Process. Technol.* **2007**, *88*, 1099–1107. [CrossRef]
28. Gupta, S.K.; Wall, T.F.; Creelman, R.A.; Gupta, R.P. Ash fusion temperatures and the transformations of coal ash particles to slag. *Fuel Process. Technol.* **1998**, *56*, 33–43. [CrossRef]
29. Lee, B.-H.; Kim, S.-G.; Song, J.-H.; Chang, Y.-J.; Jeon, C.-H. Influence of coal blending methods on unburned carbon and NO emissions in drop-tube furnace. *Energy Fuels* **2011**, *25*, 5055–5062. [CrossRef]
30. Barroso, J.; Ballester, J.; Ferrer, L.M.; Jimenez, S. Study of coal ash deposition in an entrained flow reactor: Influence of coal type, blend composition and operating conditions. *Fuel Process. Technol.* **2006**, *87*, 737–752. [CrossRef]
31. Blanchard, R. Measurements and modeling of coal ash deposition in an entrained flow reactor. Master's Thesis, Brigham Young University, Provo, UT, USA, 2008.
32. Lee, Y.-J.; Choi, J.-W.; Park, J.-H.; Namkung, H.; Song, G.-S.; Park, S.-J.; Lee, D.-W.; Kim, J.-G.; Jeon, C.-H.; Choi, Y.-C. Techno-economical method for the removal of alkali metals from agricultural residue and herbaceous biomass and its effect on slagging and fouling behavior. *ACS Sustain. Chem. Eng.* **2018**, *6*, 13056–13065. [CrossRef]
33. Wall, T.F.; Gupta, S.K.; Gupta, R.P.; Sanders, R.H.; Creelman, R.A.; Bryant, G.W. False deformation temperatures for ash fusibility associated with the conditions for ash preparation. *Fuel* **1999**, *78*, 1057–1063. [CrossRef]
34. Tortosa Masia, A.A.; Buhre, B.J.P.; Gupta, R.P.; Wall, T.F. Use of TMA to predict deposition behavior of biomass fuels. *Fuel* **2007**, *86*, 2446–2456. [CrossRef]

35. Priyanto, D.E.; Ueno, S.; Sato, N.; Kasai, H.; Tanoue, T.; Fukushima, H. Ash transformation by co-firing of coal with high ratios of woody biomass and effect on slagging propensity. *Fuel* **2016**, *174*, 172–179. [CrossRef]
36. Rong, H.; Wang, T.; Zhou, M.; Wang, H.; Hou, H.; Xue, Y. Combustion characteristics and slagging during co-combustion of rice husk and sewage sludge blends. *Energies* **2017**, *10*, 438. [CrossRef]
37. Namkung, H.; Lee, Y.-J.; Park, J.-H.; Song, G.-S.; Choi, J.-W.; Choi, Y.-C.; Park, S.-J.; Kim, J.-G. Blending effect of sewage sludge and woody biomass into coal on combustion and ash agglomeration behavior. *Fuel* **2018**, *225*, 266–276. [CrossRef]

energies

MDPI

Article

Comparative Thermogravimetric Assessment on the Combustion of Coal, Microalgae Biomass and Their Blend

Ricardo N. Coimbra [1], Carla Escapa [2] and Marta Otero [1,3,*]

1 Department of Environment and Planning, University of Aveiro, Campus Universitário de Santiago, 3810-193 Aveiro, Portugal
2 Department of Applied Chemistry and Physics, Institute of Environment, Natural Resources and Biodiversity (IMARENABIO), Universidad de León, 24071 León, Spain
3 Centre for Environmental and Marine Studies (CESAM), University of Aveiro, Campus Universitário de Santiago, 3810-193 Aveiro, Portugal
* Correspondence: marta.otero@ua.pt

Received: 5 July 2019; Accepted: 29 July 2019; Published: 1 August 2019

Abstract: In this work, thermogravimetric analysis (TGA), differential thermogravimetry (DTG), and differential scanning calorimetric (DSC) were used to assess the combustion of microalgae biomass, a bituminous coal, and their blend. Furthermore, different correlations were tested for estimating the high heating value of microalgae biomass and coal, with both materials possessing similar values. TGA evidenced differences between the combustion of the studied fuels, but no relevant interaction occurred during their co-combustion, as shown by the DTG and DSC curves. These curves also indicated that the combustion of the blend mostly resembled that of coal in terms of weight loss and heat release. Moreover, non-isothermal kinetic analysis revealed that the apparent activation energies corresponding to the combustion of the blend and coal were quite close. Overall, the obtained results indicated that co-combustion with coal might be a feasible waste to energy management option for the valorization of microalgae biomass resulting from wastewater treatment.

Keywords: sustainable energy; thermal valorization; kinetic modelling; iso-conversional methods; zero-waste water treatment; *Chlorella sorokiniana*

1. Introduction

The rapid growing of population continuously increases the global demand for energy. This demand is actually mainly satisfied by the consumption of oil followed by coal, which remain the world's leading fuels, respectively accounting for 33% and 30% of global energy consumption [1]. Such an extensive use is leading to the depletion and ever-rising prices of these fossil non-renewable fuels [2]. On the other hand, greenhouse gases emission from burning of fossil fuels, mainly CO_2, is the main cause of global warming [3]. These facts have motivated increasing research efforts regarding alternative and/or non-conventional energy resources. Among them, biomass, which may be categorized into first, second, and third generation according to its origin, has been recognized as a promising option, since it is sustainable, renewable, and less polluting [2–4].

Microalgae biomass may be considered to be a third generation biofuel, which holds several advantages, such as microalgae rapid growth rate, high oil content, high yield per area, no competition with crops for arable land or freshwater [5,6]. Furthermore, microalgae are considered as promising candidates for CO_2 bio-sequestration [7]. However, the implementation of CO_2 sequestration by microalgae is mostly limited by techno-economic constrains [8]. Microalgae may be cultivated in wastewater, allowing for simultaneous CO_2 mitigation and wastewater treatment [7]. In order to increase economic feasibility [7]. In this way, wastewater is used as source of nutrients and water, which allows for reducing the costs of microalgae culturing [5,9,10]. On the other hand, microalgae are efficient microorganisms for wastewater treatment, since they are able to remove not only nutrients, but also heavy metals [11] and emerging contaminants [12,13]. Still, during microalgae cultivation, residual biomass is generated and use should be given to this biomass within the actual circular economy context [14]. Therefore, the utilization of microalgae biomass as a third generation biofuel might be an option for closing the loop and increasing the sustainability of microalgae culture [8]. In this sense, integrating microalgae culture-wastewater treatment-biofuel production allows for carbon dioxide mitigation and wastewater treatment, while providing biofuel feedstock in a much cleaner manner [7,15].

Biomass thermochemical conversion is considered to be one of the most effective and promising routes aimed at the use of biomass for energy purposes. Thermochemical processes typically include [16]: pyrolysis, gasification and combustion, which is the most commonly used pathway to extract energy from biomass [17]. Algal biomass has lower decomposition temperatures during thermochemical conversion when compared to lignocellulosic biomass, which is due to differences in their major components and results in higher reactivity and lower operational costs [18]. In any case, a good understanding of microalgae behaviour during thermochemical conversion is essential in efficient processing.

The utilization of thermogravimetric analysis (TGA) for the characterization of thermal decomposition during the combustion of coal is well established, being more recent its use for biomassic fuels [19] and their co-processing with coal [20–22]. Such utilization is advantageous, since TGA offers a rapid evaluation of the thermal decomposition of any fuel, the initial and final temperatures of combustion, and other important features, such as maximum reactivity temperature or interaction between fuels during co-processing [19,23].

Comparatively with lignocellulosic biomass, only very recent and few works are concerned about the thermal analysis of microalgae combustion [18] and studies on the co-combustion of microalgae biomass with fossil fuels, such as coal, are even scarcer. However, co-combustion with fossil fuels is an interesting option that may help to reduce the consumption of non-renewable resources for power generation, while allowing for the utilization of existing infrastructures. Thus, in this manuscript, simultaneous TGA and differential scanning calorimetry (DSC) were used to assess the combustion behaviour of microalgae biomass and its blend with coal. The main aims were to evaluate the effect that blending with microalgae biomass has on the combustion of coal and its kinetics and to find out whether interactions between both fuels occur during their co-combustion.

2. Materials and Methods

2.1. Microalgae Culturing

For this study, *Chlorella sorokiniana* CCAP 211/8 K (UTEX Culture Collection) was cultured in axenic conditions. This strain was chosen because Chlorella sp. is amongst the most commonly used for wastewater treatment, possesses high growth rates, high light to biomass conversion, ability to grow under phototrophic, photomixotrophic and heterotrophic conditions, high protein amount, essential amino acids, and fatty acids [24,25]. On the other hand, as compared with *C. vulgaris*, the thermal decomposition of *C. sorokiniana* has been less studied in the literature.

The inoculum of *C. sorikiniana* was cultivated in 250 mL Erlenmeyer flasks in the standard culture medium Mann and Myers [26], which is composed of (per litre of distilled water): 1.2 g $MgSO_4{\cdot}7H_2O$, 1.0 g $NaNO_3$, 0.3 $CaCl_2$, 0.1 g K_2HPO_4, 3.0×10^{-2} g Na_2EDTA, 6.0×10^{-3} g H_3BO_3, 2.0×10^{-3} g $FeSO_4{\cdot}7H_2O$, 1.4×10^{-3} g $MnCl_2$, 3.3×10^{-4} g $ZnSO_4{\cdot}7H_2O$, 7.0×10^{-6} g $Co(NO_3)_2{\cdot}6H_2O$, 2.0×10^{-6} g $CuSO_4{\cdot}5H_2O$. This inoculum was grown under constant temperature (25 ± 1 °C), irradiance (175 $\mu E/m^2{\cdot}s$), photoperiod (12:12), and shaking (250 rpm) until reaching a biomass concentration of 0.1 g/L. Afterwards, the culture was grown in a 10 L-PBR after acclimatization to synthetic wastewater in 1 L-bubbling column photobioreactors (PBRs) and growing up to 0.1 g/L of biomass concentration. Each litre of synthetic wastewater was prepared by dissolving in distilled water: peptone, 160 mg; meat extract, 110 mg; urea, 30 mg; anhydrous dipotassium hydrogen phosphate (K_2HPO_4), 28 mg; sodium chloride (NaCl), 7 mg; calcium chloride dehydrate ($CaCl_2{\cdot}2H_2O$), 4 mg; and, magnesium sulphate heptahydrate ($Mg_2SO_4{\cdot}7H_2O$), 2 mg. This synthetic wastewater [27] gives a mean DOC concentration of about 100 mg/L. The culture conditions and medium used for the inoculum growth were maintained for microalgae growth in the PBRs, except for the irradiance, which was 650 $\mu E/m^2{\cdot}s$, as supplied by eight fluorescent lamps (58 W, 2150 lumen, Philips, France). Furthermore, the PBRs were aerated with filtered air (0.22 μm sterile air-venting filter, MillexFG50-Millipore), at a rate of 0.3 v/v/min., enriched with CO_2 at 7% v/v, which was injected on demand to keep a constant pH (pH = 7.5 ± 0.5), as controlled by a pH sensor. At the end of the culture, the 10 L-PBR was dismantled, and microalgae biomass was harvested by 5 min centrifugation of the cellular suspension at 6461 g in a SIGMA 2-16P centrifuge. Microalgae biomass (MB) was then washed twice with distilled water, oven dried during 24 h at 378 K, homogenized, and stored until use at 277 K.

2.2. Materials and Characterization

MB and a bituminous coal (BC) coming from the north coalfield of León (Spain) and commonly exploited in thermal power stations were used in this work. Before thermal analysis, MB and BC were grinded and sieved to have a 0.105 mm < particle diameter < 0.210 mm, which is within the size range that is commonly used in circulating fluidized boilers, allows for minimization of differences in heat of combustion values, and is large enough to ensure homogeneous ignition. Subsequently, proximate analyses of MB and BC were carried out following the procedures from ASTM D3172 to D3175 [28–31]. Elemental analysis was performed in a LECO CHNS-932, according to standard procedures, namely ASTM D5373 [32] and ASTM D4239 [33].

The high heating value (HHV) of MB and BC at a constant volume was determined by means of an isoperibol oxygen bomb calorimeter LECO AC-600 and following the procedure UNE-EN 14918:2011 [34]. Additionally, and for comparison purposes, HHV was estimated by the correlations that are listed in Table 1, together with the corresponding assumptions.

Table 1. Correlations considered for the estimation of the high heating value (HHV).

No.	Reference	CORRELATION	Originally Targeted Fuel
		Based on Elemental Analysis	
1	Dulong [35]	$HHV(MJ/kg) = -0.763 + 0.301[C] + 0.525[H] + 0.064[O]$	Biomass of any type and/or origin
2	Tillman [36]	$HHV(MJ/kg) = 0.4373[C] - 1.6701$	Biomass
3	Abe [37]	$HHV(MJ/kg) = 0.3391[C] + 1.4340[H] - 0.0970[O]$	Biomass from florestal origin
4	Demirbas et al. [38]	$HHV(MJ/kg) = 0.335[C] + 1.423[H] - 0.154[O]$	Lignocellulosic fuels
5	Sheng and Azevedo [39]	$HHV(MJ/kg) = -1.3675 + 0.3137[C] + 0.7009[H] + 0.0318[O]$	Biomass
6	Yin [40]	$HHV(MJ/kg) = 0.2949[C] + 0.8250[H]$	Lignocellulosic fuels (agricultural by-products and wood)
		Based on Proximate Analysis	
7	Jenkins and Ebeling [41]	$HHV(MJ/kg) = 26.601 - 0.304[Ash] - 0.082[VM]$	Biomass of any type and/or origin
8	Parikh et al. [42]	$HHV(MJ/kg) = 0.3536[FC] + 0.1559[VM] - 0.0078[Ash]$	Solid fuels
9	Sheng and Azevedo [39]	$HHV(MJ/kg) = -3.0368 + 0.2218[VM] + 0.2601[FC]$	Biomass
10	Majumder et al. [43]	$HHV(MJ/kg) = -10.81408 + 0.3133([VM] + [FC])$	Coal
11	Yin [40]	$HHV(MJ/kg) = 0.1905[VM] + 0.2521[FC]$	Lignocellulosic fuels (agricultural by-products and wood)
		Based on both Elemental and Proximate Analysis	

Table 1. *Cont.*

No.	Reference	CORRELATION	Originally Targeted Fuel
12	Grabosky and Bain [44]	$HHV(MJ/kg) = 0.328[C] + 1.4306[H] - 0.0237[N] + 0.0929[S] - (1 - [Ash]/100)(40.11[H]/[C])$	Biomass
13	IGT [45]	$HHV(MJ/kg) = 0.341[C] + 1.323[H] + 0.068[S] - 0.0153[Ash] - 0.1194([O] - [N])$	Coal
14	Channiwala and Parikh [46]	$HHV(MJ/kg) = 0.3491[C] + 1.1783[H] + 0.1005[S] - 0.1034[O] - 0.0151[N] - 0.0211[Ash]$	Solid, liquid and gaseous fuels
15	Sajdak et al. [47]	$HHV(MJ/kg) = 0.001 \times (601.95 - 11.57[Ash] - 7.12[VM] + 341.67[C] + 1165.86[H] - 97.35[O] - 193.37[N] + 110.36[S])$	Biomass, biochar and coal

HHV: high heating value; [C]: carbon content; [H]: hydrogen content; [N]: nitrogen content; [S]: sulphur content; [O]: oxygen content; [FC]: fixed-carbon content; [VM]: volatile matter content; [Ash]: ash content; note: all values are expressed in wt% on dry basis.

2.3. Thermal Analyses

A Setaram equipment, model SETSYS Evolution was used in this work. Non-isothermal combustion runs were carried out after calibration for baseline, weight, temperature, and heat flow. Throughout these runs, Thermogravimetry (TG) and Differential Scanning Calorimetry (DSC) signals were simultaneously registered during the temperature-programmed combustion of MB, BC, and their blend (MB-BC). In the blend, a 10% wt. of MB was used, since it has been shown that such a percentage is adequate for the practical implementation of co-combustion of coal with biomass in existing infrastructures, namely in thermal power plants [20]. Derivative TG (DTG) curves were also determined as the first derivation of TG results with respect to time. The runs were carried out up to 1200 K at four different heating rates ($\beta = dT/dt$): 0.1, 0.2, 0.4, and 0.5 K/s, in order to determine the corresponding TG-DSC curves. For each sample and β, three repetitive runs were carried out using 15 ± 1 mg of MB, BC, or MB-BC, after having verified that this mass ensured representativeness and avoided heat and/or mass transfer limitations. All the runs were done under a continuous air flow (100 cm^3/min at 1 atm of gauge pressure).

The theoretical DTG curves ($DTG_{(T)}$) and DSC curves ($DSC_{(T)}$) were calculated for the blend MB-BC using Equations (1) and (2), respectively, as a weighted average of its composition in order to check interaction between MB and BC during their co-combustion:

$$DTG_{(T)} = 0.1 \times DTG_{MB} + 0.9 \times DTG_{BC} \quad (1)$$

where DTG_{MB} and DTG_{BC} are the weight loss rate of MB and BC throughout their respective temperature-programmed combustions.

$$DSC_{(T)} = 0.1 \times DSC_{MB} + 0.9 \times DSC_{BC} \quad (2)$$

where DSC_{MB} and DSC_{BC} are the differential scanning calorimetry results that correspond to the temperature-programmed combustion of MB and BC, respectively.

2.4. Non-Isothermal Kinetic Analysis

The rate of heterogeneous solid state reactions is generally described by the following equation:

$$\frac{d\alpha}{dt} = k(T)f(\alpha) \quad (3)$$

where α is the extent of reaction or fractional conversion, t is time, T is temperature, $k(T)$ is the temperature-dependent constant, and $f(\alpha)$ is a function that describes the dependence of the reaction rate on α.

Usually, the decomposition from a solid state is mathematically described in terms of the kinetic triplet (apparent activation energy (E), pre-exponential factor (A), and an expression of the kinetics in terms of $f(\alpha)$), which may be correlated to the obtained results by this rate expression:

$$\frac{d\alpha}{dt} = Ae^{-E/RT} f(\alpha) \tag{4}$$

The previous rate expression (Equation (4)) may be transformed into a non-isothermal one, which defines the reaction rate in function of T at a constant β:

$$\frac{d\alpha}{dT} = \frac{A}{\beta} Ae^{-E/RT} f(\alpha) \tag{5}$$

When Equation (5) is integrated up to α, results in:

$$\int_0^\alpha \frac{d\alpha}{f(\alpha)} = g(\alpha) = \frac{A}{\beta} \int_{T_0}^{T} e^{-E/RT} dT \tag{6}$$

where $g(\alpha)$ is the integral reaction model.

Several methods may be used in order to obtain a description of the combustion process in terms of E [48]. These methods can be classified depending on the experimental conditions and on the mathematical analysis that was carried out. As for the experimentation, the results may be obtained either under isothermal or non-isothermal conditions. Regarding mathematical analysis, either the model-fitting or the iso-conversional (model-free) approaches may be followed.

For non-isothermal kinetic analysis, different iso-conversional models that involve carrying out temperature-programmed runs at different β [48] have been developed to determine E. During the last decade, these methods have been frequently used to study the thermal decomposition kinetics of very different types of biofuels [20,21], including microalgae biomass [49].

In this sense, E may be estimated by applying the iso-conversional model developed by Flynn, Wall, and Ozawa [50,51], which is an integral method that uses the Doyle's approximation [52]:

$$\ln(\beta) = ln\left[\frac{AE}{R\,g(\alpha)}\right] - 5.331 - 1.052\frac{E}{RT} \tag{7}$$

The utilization of the Flynn-Wall-Ozawa (FWO) method [50,51] requires the determination of the T corresponding to fixed values of α from runs carried out at different β. E is estimated by plotting $\ln(\beta)$ vs. $1/T$ for each α, which gives straight lines with slope—E/R.

E may be also determined on the basis of the Kissinger-Akahira-Sunose (KAS) kinetic model [53,54], as it is next described.

In Equation (5) $E/2RT$ » 1, therefore, the integral can be approximated by:

$$\int_{T_0}^{T} e^{-E/RT} dT \approx \frac{R}{E} T^2 e^{-E/RT} \tag{8}$$

Substituting the temperature integral and taking the logarithm:

$$\ln\frac{\beta}{T^2} - \ln\left[\frac{RA}{Eg(\alpha)}\right] - \frac{E}{R}\frac{1}{T} \tag{9}$$

For the application of the KAS model [53,54], it is necessary to carry out runs at different β, the respective conversion curves being evaluated from the measured TG curves. For each α, $\ln(\beta/T^2)$ plotted versus $1/T$ gives a straight line with slope $-E/R$.

3. Results and Discussion

3.1. Materials Characterization

The results from the proximate and elemental analyses of MB and BC are depicted in Table 2, together with the measured HHV for each material.

Table 2. Proximate analysis, elemental analysis, and calorific values for the microalgae biomass (MB) and the bituminous coal (BC) used in this work.

Properties	MB	BC
Proximate Analysis (wt. %)		
Moisture	10.1	0.8
Volatiles (d.b.)	78.2	8.2
Ashes (d.b.)	6.2	31.1
FC* (d.b.)	15.6	60.7
Elemental Analysis (wt. %, d.b.)		
C	52.0	62.7
H	6.8	2.5
N	10.7	1.3
S	0.6	0.7
O*	29.8	1.7
Calorific Analysis (MJ/kg, d.b.)		
HHV	22.9	24.3

FC: fixed-carbon; HHV: high heating value; d.b.: dry basis; * calculated by difference.

As may be seen in Table 2, the proximate and elemental analyses of MB and BC evidence that these fuels have very distinct properties due to their different origin and nature. Within the proximate analysis, moisture percentages for both materials are usual equilibrium values for storeroom conditions. With respect to the ash yield, it is quite smaller for MB (6.2%) than for BC (31.1%), which is a positive fact for the biofuel utilization of MB, since relatively high ash contents are undesirable in many combustion facilities. However, the amount of volatiles in MB (78.2%) is much larger than in BC (8.2%). The higher volatile matter content of biomass, as compared with coal, is known to improve the combustion of the latter, which results in a better burn out and lower unburned carbon in the ashes [21]. Still, adaptations of the combustor may be necessary for the co-processing of fuels with very different volatiles content. On the other hand, due to the higher volatile content of MB, it possesses a lower fixed-carbon (15.6%) than BC (60.7%), which is expected to be corroborated in their separate combustion DTG profiles.

Regarding the elemental analysis, BC has a higher C content (62.7%) than MB (52.0%). Contrarily, BC has much lower O content (1.7%) than MB (29.8%). Additionally, the H and N contents of BC (2.5 and 1.3%, respectively) are lower than those of MB (6.8 and 10.7%, respectively). Nonetheless, it has been demonstrated that NO emissions are not strongly dependent on the fuel nitrogen content [55]. Meanwhile, both BC and MB have a similar S content (0.7 and 0.6%, respectively), so their combustion may involve similar SO_x emissions.

The results from proximate and elemental analyses of MB are very similar to those that were determined for *C. vulgaris* biomass by Gao et al. [56]. However, slightly different results have been obtained by other authors for *C. vulgaris* biomass [57], being especially relevant the comparatively higher volatiles content and lower percentage of ashes of MB. The differences are probably related to the specific strain and the way or stage of culturing. In fact, nitrogen supplementation [58] and the age of the culture [59] have already been shown to affect thermal properties of *C. sorokiniana*. Regarding BC, the results are comparable to previously published data for coal with the same rank and origin [23].

Finally, as regards the HHV, both MB and BC have very similar values (22.9 and 24.3 MJ/kg, respectively), which means that the combustion of their blend is not going to have remarkable energetic effects as compared with the combustion of BC. The HHV determined in this work for MB is within the range of values referred in the review by Chen et al. [4], namely 14 to 24 MJ/kg, while the HHV of BC is close to published data for coal with the same rank and origin [23].

As for the estimation of HHV, Figure 1 shows the values that were obtained by the correlations depicted in Table 1 for each MB and BC, together with the measured value and a deviation limit of ±10%. Some of the correlations satisfactorily estimate the HHV of MB and BC, mainly those that are based on elemental or on both elemental and proximate analyses. However, correlations that are only based on proximate analysis (No. 7–11 in Table 1) mostly underestimate the HHV, except for correlation No. 10 [43], which overestimates the HHV of MB, and correlations No. 8 [42] and 10 [43], which only give an acceptable HHV estimation for BC. Correlation No. 1, which is the well-known Dulong correlation [35] and is just based on the C, H, and O contents, correlation No. 4 [38], which also depends on the C, H and O contents, and correlation No. 14 [46], which stands on the elemental analysis and the ash content, are those that more closely estimate the HHV for both MB and BC. These correlations may be very useful for the quick estimation of the calorific potential of microalgae and coal when planning their co-processing.

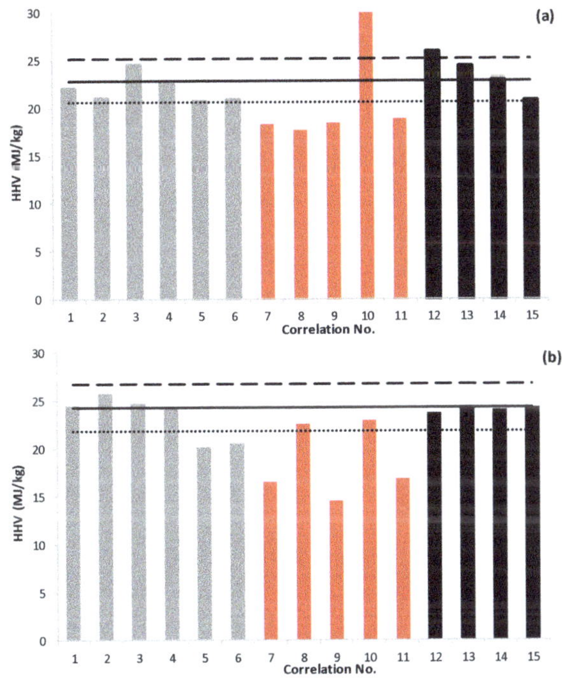

Figure 1. Estimated High Heating Value (HHV) for microalgae biomass (MB) (**a**) and BC (**b**) by correlations No. 1 to No. 15 listed in Table 1. For each material, the estimated HHV values are represented by verticals bars, the measured value is represented by a continuous horizontal line, the upper limit (measured HHV + 10%) is represented by a dashed line, and the lower limit (measured HHV − 10%) is marked with a dotted line.

To the best of our knowledge, there are no published studies using correlations for the estimation of HHV in the specific case of microalgae biomass. Therefore, the applicability of the here considered correlations (Table 1) was tested in this work for available data in the literature on the calorific value of microalgae biomass from different strains. In the case of those microalgae biomasses for which just the elemental analysis is available, the estimated HHV values are depicted in Table 3. For biomasses whose elemental and proximate analyses are available, the estimations of HHV are displayed in Table 4.

Table 3. Published results on the elemental and calorific analysis for different microalgae biomasses. For each case, the published High Heating Value (HHV) is shown together with the HHV estimated by correlations based on elemental analysis displayed in Table 1. Estimated HHVs that are within ± 10% the measured value are in bold.

	Chlorella	*Chlorella vulgaris*	*Chlorella vulgaris*	*Chlorella vulgaris*	*Chlorella vulgaris*	*Chlorella vulgaris* **residue**	*Chlamydomonas reinhardtii*	*Chlamydomonas reinhardtii*	*Dunaliella tertiolecta*	*Nannocloropsis oceanica*	*Nannocloropsis oceanica* **residue**	*Spirulina platensis*	*Spirulina platensis*	*Scenedesmus obliquus*
References	Babich et al. [60]	Xu et al. [61]	Xu et al. [61]	Wang et al. [62]	Kebelmann et al. [63]	Wang et al. [62]	Kebelmann et al. [63]	Kebelmann et al. [63]	Zou et al. [64]	Chen et al. [65]	Chen et al. [65]	Jena and Das [66]	Wu et al. [67]	Chen et al. [68]
Elemental Analysis (wt. %, d.b.)														
C	50.2	45.8	53.8	42.51	43.9	45.04	52	50.2	39	50.06	45.24	46.16	45.7	37.37
H	7.3	5.6	7.72	6.77	6.2	6.88	7.4	7.3	5.37	7.46	6.55	7.14	7.71	5.8
N	9.3	4.6	1.1	6.64	6.7	9.79	10.7	11.1	1.99	7.54	11.07	10.56	11.26	6.82
S	-	-	-	-	-	-	-	-	0.62	0.47	0.56	0.74	0.75	-
O	33.2	38.7	37	27.95	43.3	29.42	29.8	31.4	53.2	34.47	36.58	35.44	25.69	50.02
Calorific Analysis (MJ kg^{-1}, d.b.)														
HHV (measured)	21.2	18.4	24.0	16.8	18.0	19.4	23.0	22.0	14.2	21.5	18.2	20.5	20.5	16.1
HHV (No. 1)	**21.5**	**16.6**	**22.7**	19.1	16.0	**19.9**	**22.9**	**21.9**	11.3	**21.5**	**18.2**	**19.5**	**22.0**	12.0
HHV (No. 2)	**20.3**	**18.4**	**21.9**	**16.9**	**17.5**	**18.0**	**21.1**	**20.3**	**15.4**	**20.2**	**18.1**	**18.5**	**18.3**	**14.7**
HHV (No. 3)	24.2	**19.8**	**25.7**	21.4	**19.6**	22.3	25.4	24.4	15.8	24.3	21.2	**22.5**	24.1	**16.1**
HHV (No. 4)	**22.0**	**17.4**	**23.3**	19.6	**16.9**	**20.3**	**23.4**	**22.4**	**12.5**	**22.1**	**18.8**	**20.2**	**22.3**	13.1
HHV (No. 5)	**20.5**	**18.2**	**22.1**	**17.6**	**18.1**	**18.5**	**21.1**	**20.5**	16.3	**20.7**	**18.6**	**19.2**	**19.2**	**16.0**
HHV (No. 6)	**20.8**	**18.1**	**22.2**	**18.1**	**18.1**	**19.0**	**21.4**	**20.8**	15.9	**20.9**	**18.7**	**19.5**	**19.8**	**15.8**

Table 4. Published results on the proximate, elemental and calorific analysis for different microalgae biomasses. For each case, the published High Heating Value (HHV) is shown together with the HHV estimated by correlations displayed in Table 1. Estimated HHVs that are within ± 10% the measured value are in bold.

	Chlamydomonas	*Chlorella sorokiniana*	*Chlorella sorokiniana*	*Chlorella vulgaris*	*Chlorella vulgaris*	*Isochrysis galbana*	*Nannochloropsis limnetica*	*Nannochloropsis gaditana*	*Phaeodactylum tricornutum*	*Spirulina platensis*	*Scenedesmus almeriensis*
References	Bui et al. [69]	Bui et al. [69]	Paniagua et al. [59]	Chen et al. [70]	Soria-Verdugo et al. [71]	Soria-Verdugo et al. [71]	Soria-Verdugo et al. [71]	Soria-Verdugo et al. [71]	Soria-Verdugo et al. [71]	Soria-Verdugo et al. [71]	López et al. [72]
Proximate Analysis (wt. %, d.b., except for moisture (wt. %))											
Moisture	3.5	3.8	9.6	-	-	-	-	-	-	-	5.4
Volatiles	75.5	73.2	76.1	55.37	76.26	86.13	84.06	81.56	62.1	81.46	73.1
Ashes	5.2	7.9	7.83	10.28	13.11	8.31	10.52	9.16	25.46	6.4	20
FC	15.6	15.1	16.07	34.35	10.63	5.56	5.42	9.28	12.44	12.14	6.9
Elemental Analysis (wt. %, d.b.)											
C	40.32	45.07	47.9	47.84	51.317	43.644	52.453	52.805	40.647	49.720	43.84
H	7.38	7.64	6.4	6.41	7.655	6.620	8.062	7.803	6.612	7.338	6.08
N	2.61	3.88	8.74	9.01	9.897	5.474	7.883	8.230	6.813	11.550	6.8
S	-	-	0.78	1.46	0.573	0.816	0.617	0.509	1.446	0.693	0.32
O	44.5	35.52	36.18	25	17.448	35.136	20.464	21.493	19.023	24.299	22.96
Calorific Analysis (MJ kg^{-1}, d.b.)											
HHV (measured value)	17.41	20.4	18.7	21.9	22.9	19.97	23.51	24.5	19.34	22.62	20.91
HHV (No. 1)	**16.3**	**19.9**	**18.9**	**20.9**	**25.3**	**18.0**	**25.7**	**25.2**	**19.9**	**23.0**	**19.5**
HHV (No. 2)	**16.0**	**18.0**	**19.3**	19.3	**20.8**	17.4	**21.3**	21.4	16.1	20.1	17.5
HHV (No. 3)	19.9	22.8	21.9	**23.0**	26.7	**20.9**	27.4	**27.0**	21.4	25.0	**21.4**
HHV (No. 4)	**17.2**	**20.5**	**19.6**	**21.3**	25.4	**18.6**	**25.9**	**25.5**	**20.1**	**23.4**	**19.8**
HHV (No. 5)	**17.9**	**19.3**	**19.3**	18.9	**20.7**	**18.1**	**21.4**	21.4	**16.6**	20.1	17.4
HHV (No. 6)	**18.0**	**19.6**	**19.4**	19.4	**21.4**	**18.3**	**22.1**	22.0	**17.4**	**20.7**	17.9
HHV (No. 7)	**18.8**	**18.2**	**18.0**	18.9	16.4	17.0	16.5	17.1	13.8	18.0	14.5
HHV (No. 8)	**17.2**	16.7	**17.5**	**20.7**	15.5	15.3	14.9	15.9	13.9	16.9	13.7
HHV (No. 9)	**17.8**	17.1	**18.0**	18.2	16.6	17.5	17.0	17.5	14.0	18.2	15.0
HHV (No. 10)	29.8	28.8	29.4	30.0	28.5	30.1	29.3	29.9	24.1	30.9	25.3
HHV (No. 11)	**18.3**	17.8	**18.5**	19.2	17.2	17.8	17.4	17.9	15.0	18.6	15.7
HHV (No. 12)	-	-	24.2	24.3	27.4	23.2	28.3	27.9	23.4	26.0	23.1
HHV (No. 13)	-	-	21.5	**22.8**	26.6	**20.0**	26.9	**26.6**	**20.9**	25.1	**20.8**
HHV (No. 14)	-	-	**20.3**	**21.5**	**24.8**	**19.2**	**25.4**	**25.1**	**19.5**	**23.3**	**19.6**
HHV (No. 15)	-	-	**18.7**	**19.9**	**22.8**	**18.1**	**23.8**	**23.4**	**18.5**	**21.0**	18.4

As may be observed in Table 4, estimations that are just based on proximate analyses (correlations No. 7 to 11 in Table 1) are mostly inadequate, while those that are just based on the elemental analysis (Table 3) or elemental analysis together with proximate analysis (Table 4) give estimations that are more satisfactory. On the whole, and coincidently with the case of microalgae biomass that was obtained in this work, correlations No. 1, 4, and, especially, No. 14 were the most accurate for estimating the HHV of microalgae biomasses in the literature.

3.2. Thermal Analysis

The TG curves that were obtained from the temperature programmed combustion of MB, BC, and their blend MB-BC at different β (0.1, 0.2, 0.4, and 0.5 K/s) are depicted in Figure S1. The resultant DTG curves are shown in Figure 2 together with the weighted calculated curves (Equation (1)) corresponding to the combustion of MB-BC. As for a typical combustion profile, mass loss occurs along with increasing temperature under oxidizing atmosphere until the volatiles and fuel content of the sample is exhausted, and then the mass of ashes remains stable. In the case of BC, the loss of volatiles and char gasification occur in a single step due to the large contribution of fixed carbon (see Table 2) to the mass loss during combustion. On the other hand, a slight weight gain that is related to oxygen chemisorption may be observed during the combustion of BC but is not present in the DTG curves corresponding to MB. Differently from BC, the DTG curve that corresponds to MB shows that mass loss occurs in two main stages, which has already observed by other authors for the combustion of microalgae [16,56,57]. The first stage, which is the most remarkable, is attributed to the devolatilization of MB and volatiles combustion and extends until 670 K. It is to highlight that this first stage occurs in a temperature range for which no mass loss is observed for BC. This is due to the higher volatiles/fixed-carbon ratio of MB (5.01), as compared to that of BC (0.13), which involves microalgae combustion predominantly occurring in gas-phase due to the combustion of volatiles. Above 670 K and ending at around 1000 K, occurs the second stage, which comprises three subsequent steps at $\beta = 0.1$ and 0.2 K/s that overlap at higher β. Despite evident differences between MB and BC, DTG experimental curves corresponding to the combustion of MB-BC mostly resemble those of BC, except for the mass gain associated to oxygen chemisorption, which is roughly appreciable. This is also true for the weighted calculated curves MB-BC (T), since they are nearly coincident with the experimental curves. This fact indicates that interaction between MB and BC during combustion is not relevant, which is favourable in terms of the practical application of co-processing. A good correlation between experimental and weight calculated DTG curves was also observed by Gao et al. [56] for the co-combustion of a lignite coal and microalgae biomass from *Chlorella vulgaris* (blending ratio 50%). These authors stated that the synergetic effects in the co-combustion of these materials were negligible [56]. Differently, and after having observed inhibitive effects during co-pyrolysis of a sub-bituminous coal that was blended with *Chlorella vulgaris* biomass (blending ratios 30, 50 and 70%) [73], Chen et al. [74] verified that, although no synergetic effects occurred in the initial and final stages of their co-combustion, some interaction occurred at the intermediate stage. Interactions were especially evident for the highest ratio of microalgae within the blend (70%) and at relative high temperatures, which has been also observed for the co-combustion of coal with composite biomass pellets that were made from catkins, wood waste, and rice straw [75]. Interactions between coal and biomass during co-combustion reported in the literature have been attributed to the acceleration of coal devolatilization during co-combustion, due to the reaction with the active radicals that were produced during biomass devolatilization [56]. Such acceleration may be ascribed to the higher temperature of coal particle surface during its co-combustion with biomass than during its individual combustion.

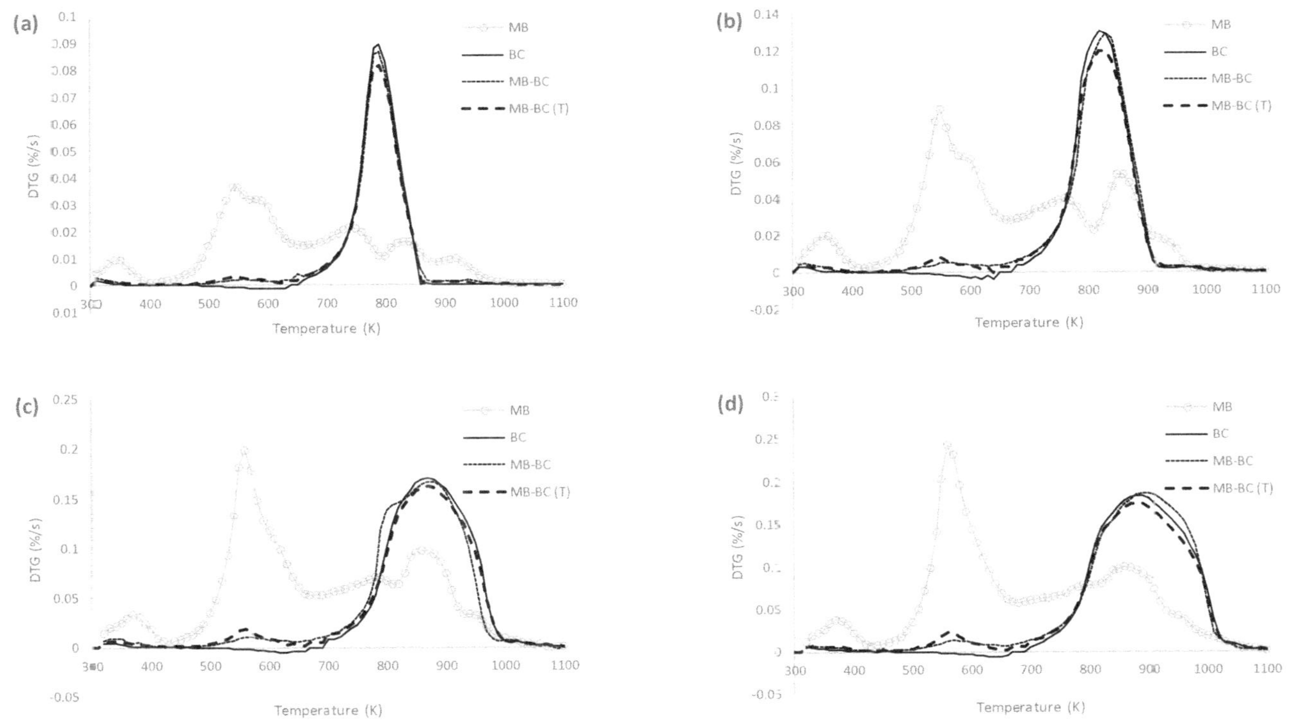

Figure 2. Differential thermogravimetry (DTG) curves corresponding to the temperature programmed combustion of MB, BC and their blend (MB-BC) together with the weighted calculated curve corresponding to the blend (MB-BC (T)) at the different heating rates here used, namely 0.1 K/s (**a**), 0.2 K/s (**b**), 0.4 K/s (**c**), and 0.5 K/s (**d**).

Figure 2 evidences that DTG curves that correspond to MB and BC were affected by increasing the heating rate (β). In this sense, it is observable an increase of the temperature at which mass loss begins; a broader range of temperature in which mass loss occurs; higher weight loss (%/s) peaks; and overlapping of sub-steps in the combustion of MB. Still, the DTG curves of the blends remain analogous to those of BC, even at increasing heating rates (β).

Table 5 depicts the characteristic parameters of the DTG combustion profiles. These parameters confirm the differences between MB and BC that were observed in Figure 2 and similitudes between BC and MB-BC. In this sense, the T_v and T_f in Table 5 correspond to the temperatures at which mass loss begins and ends, respectively. Meanwhile, for each DTG curve, the T_m corresponds to the temperature at which the maximum mass loss rate occurs, that is, the DTG_{max}. For all MB, BC and MB-BC, the temperatures T_v, T_f and T_m increase with β. For MB, the T_v and T_m values are around 240 K lower than for BC. Differently, T_v values for MB-BC are just slightly lower than for BC, due to the absence of chemisorption mass loss in the combustion of MB-BC, while T_m values are mostly coincident. On the other hand, lower T_f values are observed for BC than for MB-BC, which shows slightly lower values than MB. Finally, at β = 0.1 and 0.2 K/s, the DTG_{max} corresponding to MB are the lowest, while at β = 0.4 and 0.5 K/s they are the highest, which evidences the intensification of devolatilization of MB with β. Meanwhile, at each β, BC and MB-BC show very close DTG_{max}.

Table 5. Characteristic parameters of DTG combustion curves determined for microalgae biomass (MB), bituminous coal (BC) their blend (MB-CB).

	B (K/s)	T_v (K)	T_m (K)	T_f (K)	DTG_{max} (%/s)
MB	0.1	400	542	1031	0.0363
	0.2	410	543	1040	0.0839
	0.4	433	554	1100	0.1981
	0.5	440	557	1120	0.2457
BC	0.1	640	782	874	0.0890
	0.2	657	820	930	0.1295
	0.4	671	867	1035	0.1669
	0.5	675	875	1050	0.1820
MB-BC	0.1	600	781	981	0.0873
	0.2	631	823	993	0.1281
	0.4	646	870	1046	0.1645
	0.5	654	885	1080	0.1844

T_v: onset temperature for volatile release and mass loss; T_m: temperature of maximum mass loss rate; T_f: final combustion temperature detected as mass stabilization; DTG_{max}: maximum mass loss rate.

Figure 3 represents the DSC curves that were obtained from the temperature programmed combustion of MB, BC, and their blend MB-BC at different β (0.1, 0.2, 0.4, and 0.5 K/s), together with the weighted calculated curves (Equation (2)) corresponding to the combustion of the blend. As it may be seen, heat release during BC combustion occurs in two stages, namely during the chemisorption mass gain and, especially, during the fixed-carbon combustion that was observed in Figure 2.

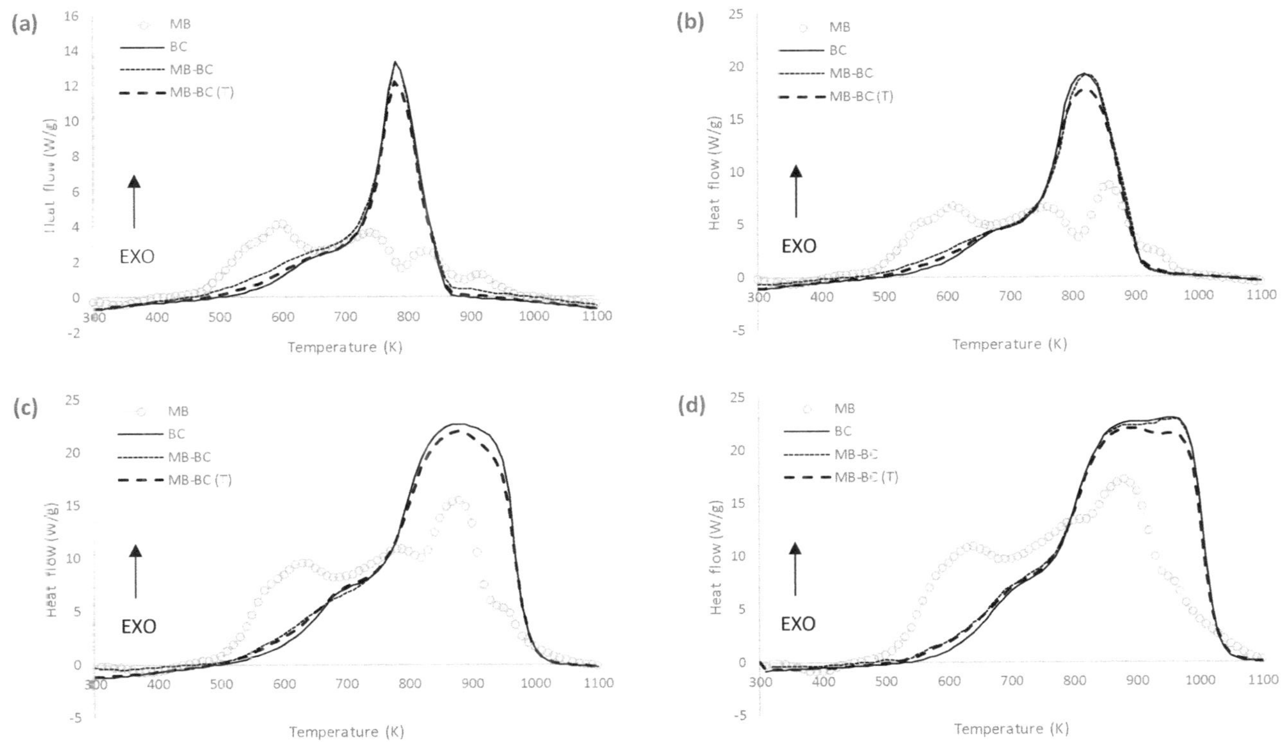

Figure 3. Differential scanning calorimetric (DSC) curves corresponding to the temperature programmed combustion of MB, BC and their blend (MB-BC) together with the weighted calculated curve corresponding to the blend (MB-BC (T)) at the different heating rates here used, namely 0.1 K/s (**a**), 0.2 K/s (**b**), 0.4 K/s (**c**) and 0.5 K/s (**d**).

Heat liberation during MB combustion takes place at four subsequent steps, corresponding to the combustion of volatiles, organic material, fixed-carbon, and char, which are evident at β = 0.1 K/s and progressively overlap at increasing β. Consequently, at β = 0.5 K/s, heat release during MB combustion occurs just in two main stages. The second one, which corresponds to the combustion of the fixed-carbon content, is centred at the same temperature than the second stage of BC (Figure 3). Although DSC analysis helps to obtain a more realistic approach to the combustion process of biomass, there are few published DSC results regarding the combustion of microalgae biomass to compare with the here obtained. Yet, López-González et al. [16] also observed two heat release stages during the temperature programmed combustion of biomass from three different microalgae strains at β = 0.67 K/s. Regarding the combustion of MB-BC in this work, it may be observed in Figure 3 that, even when DSC curves corresponding to MB and BC are very different, heat release during the combustion of MB-BC shows the same trend than BC.

The characteristic temperatures at DSC curves in Figure 3 are depicted in Table 6, together with the enthalpies of combustion, which were calculated for each MB, BC, and MB-BC by the integration of the corresponding exothermic peak. The onset temperatures for heat release (T_i) are slightly lower for MB than for BC, with the latter being close to those of MB-BC. Contrarily, the final combustion temperatures (T_e) are higher for MB than for BC, which are again very close to those that were observed for MB-BC. Regarding temperatures of maximum energy release (T_{max}), the lowest value is observed for MB combustion at β = 0.1 K/s, since the volatiles combustion was the main peak of heat release. However, at increasing β, the third stage (corresponding to the fixed-carbon combustion) progressively gains more prominence, so higher T_{max} values occur. At each β, T_{max} values that were observed for BC and MB-BC are equivalent and correspond to the fixed-carbon combustion peak. Finally, lower enthalpy (ΔH) values are obtained for MB combustion than for BC and MB-BC. In any case, the ΔH here obtained for MB are slightly higher than those that were determined by López-González et al. [16], which were within 7.8 and 8.8 kJ/g.

Table 6. Characteristic parameters of DSC combustion curves determined for microalgae biomass (MB), bituminous coal (BC) their blend (MB-CB).

	β (K/s)	T_i (K)	T_{max} (K)	T_e (K)	ΔH (kJ/g)
MB	0.1	450	597	1000	11.62
	0.2	460	857	1019	11.88
	0.4	470	876	1100	11.76
	0.5	475	879	1165	11.71
BC	0.1	500	782	893	13.87
	0.2	500	820	968	13.81
	0.4	500	880	1042	13.96
	0.5	500	959	1086	13.78
MB-BC	0.1	461	782	950	14.16
	0.2	470	825	980	14.12
	0.4	475	875	1046	14.15
	0.5	480	964	1080	14.11

T_i: onset temperature for energy release and peak integration; T_{max}: temperature of maximum energy release during combustion; T_e: temperature at the end of energy release during combustion; ΔH: enthalpy determined by integration of the heat release peak in the corresponding DSC combustion profile.

3.3. Non-Isothermal Kinetic Analysis

The TG curves that correspond to the temperature programmed combustions of MB, BC, and MB-BC at the different β here considered are represented in Figure 4. For each case, six different percentages of conversion are pointed out in each curve: 10, 20, 30, 40, 50, and 60%. As it was previously observed in DTG curves (Figure 2), mass loss at relatively low temperatures was more remarkable for MB than for BC due to the high volatiles content of the first (Table 2). On the other hand, the ash yield, or residual mass after burning, was 6, 31, and 28% for MB, BC, and MB-BC. These yields are in agreement with ash contents in Table 2, and, in the case of MB-BC with the weighted calculated value for the blend.

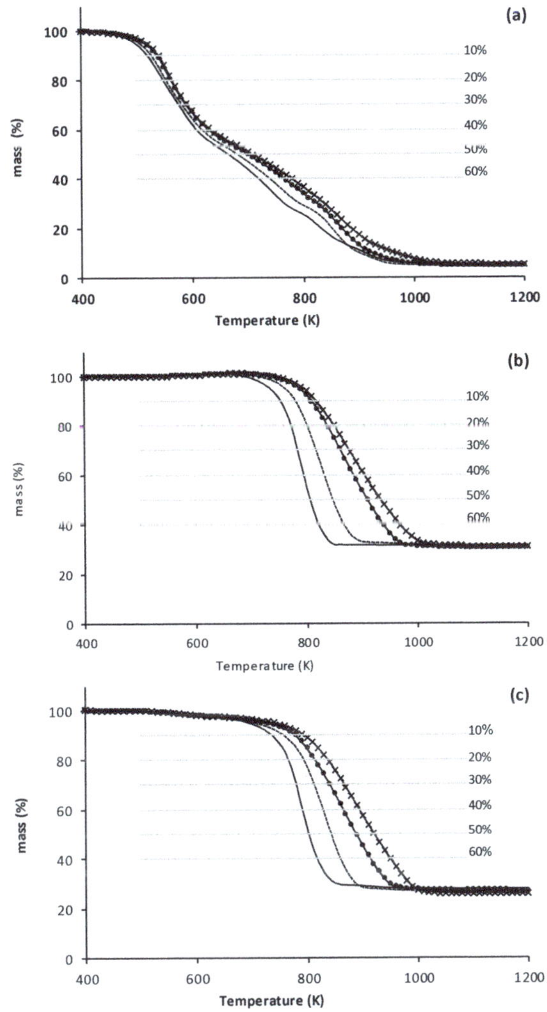

Figure 4. Thermogravimetry (TG) curves corresponding to the combustion of MB (**a**), BC (**b**), and MB-BC (**c**) at different heating rates. Conversion percentages (10, 20, 30, 40, 50, and 60%) have been marked by straight lines crossing experimental data.

Figure 5 shows the plots of ln (β/T) and ln (β/T^2) vs. $1/T$ to the several conversion degrees (α), corresponding to the combustion of MB, BC, and MB-BC, as for the respective application of the FWO model [50,51] and the KAS model [53,54]. Meanwhile, Table 7 depicts the corresponding estimations of E.

The E values estimated by the FWO model are slightly higher than those that were estimated by the KAS one, except for MB, with both models giving nearly coincident values. As it may be seen in Table 7, the average E values (E^*) that were determined for the combustion of MB are more than double of that corresponding to BC. On the other hand, the E^* values corresponding to the combustion of MB-BC are just slightly higher than what would proportionally correspond for the relative MB and BC contents (100 and 91 kJ/mol, for the FWO and KAS models, respectively).

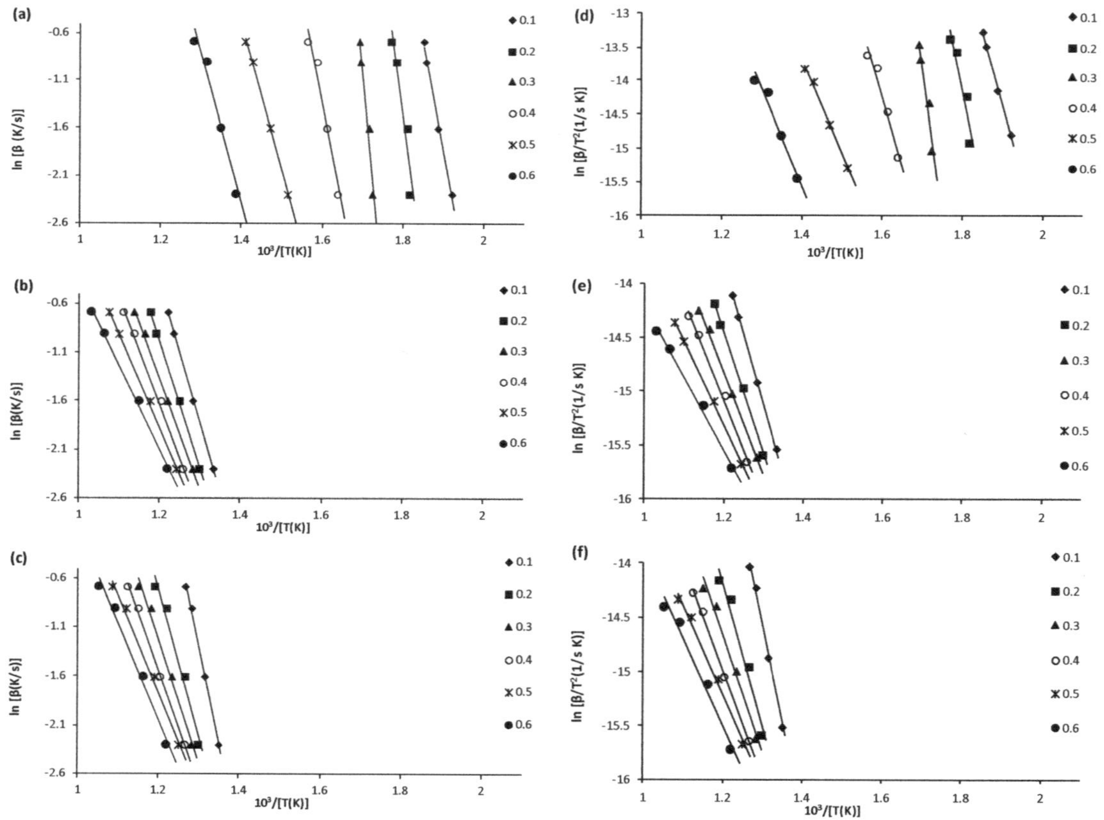

Figure 5. Results on the combustion of MB (**a**,**d**), BC (**b**,**e**) and MB-BC (**c**,**f**) at the different heating rates together with fittings to the iso-conversional kinetic models of FWO (**a**–**c**) and KAS (**d**–**f**). Note: The *Y*-axis of graphs was adjusted for the correct visualization of results.

Table 7. Values of apparent activation energy (E, kJ/mol) determined for the combustion of MB, BC and their blend (MB-BC) at the considered conversion degrees (α), together with R^2 corresponding to fittings in Figure 5.

	α	E (kJ/mol) FWO	R^2	E (kJ/mol) KAS	R^2
MB	0.1	177	0.9995	177	0.9954
	0.2	243	0.9978	246	0.8958
	0.3	343	0.9882	351	0.9322
	0.4	171	0.9953	169	0.9587
	0.5	121	0.9864	116	0.9971
	0.6	123	0.9844	117	0.9588
		197* ± 85		196* ± 90	
BC	0.1	112	0.9995	105	0.9950
	0.2	102	0.9978	94	0.9971
	0.3	89	0.9882	80	0.9973
	0.4	86	0.9953	77	0.9926
	0.5	76	0.9864	65	0.9974
	0.6	67	0.9844	56	0.9925
		89* ± 16		79* ± 18	
MB-BC	0.1	152	0.9995	147	0.9970
	0.2	118	0.9978	111	0.9661
	0.3	98	0.9882	90	0.9784
	0.4	91	0.9953	82	0.9973
	0.5	80	0.9864	68	0.9896
	0.6	77	0.9844	67	0.9762
		103* ± 28		94* ± 30	

E^* ± standard deviation from E values at the considered α.

In Table 7, the change of E with α, which is especially evident for MB, may be related to the complexity of the fuel composition and subsequently to the complex reactions ongoing under combustion, as it has already been observed for woody biomass [76]. The E^* values here determined for BC are close to those previously obtained for a coal of the same rank and origin [23], while those of MB are in consonance with published results on the combustion of microalgae biomasses. Using the FWA and the KAS methods, Chen et al. [70] determined E^* values between 134 and 242 kJ/mol for the combustion of *C. vulgaris* under 20%O_2/80%N_2 and 80%O_2/20%N_2, respectively. Applying a modification of the FWO method, namely the Starink's model, Zhao et al. [77] determined E^* values between 93 and 142 kJ/mol for the combustion of different microalgae strains, being 118 kJ/mol the E^* corresponding to *C. sorokiniana* C74. For the oxy-fuel co-combustion of *C. vulgaris* biomass with lignite and using the FWO and the KAS methods, Gao et al. [56] determined E^* values for the blends that were between 150 kJ/mol (lignite/microalgae = 7:3) and 197 kJ/mol (lignite/microalgae = 3:7) under air atmosphere (21%O_2/79%N_2). These authors [56] highlighted that the lowest and highest E^* were obtained for lignite coal and *C. vulgaris* microalgae (respectively, 146 and 213 kJ/mol), and that a higher percentage of microalgae in the blend resulted in a higher E^*. On the other hand, Chen et al. [74] determined E^* values of 68 and 107 kJ/mol for the combustion of a sub-bituminous coal and *C. vulgaris* biomass, respectively, by the FWO method and of 57 and 103 kJ/mol by the KAS one. Subsequently, these authors [74] found that with the increasing content of *C. vulgaris* in the blends, the E^* first decreased and then increased, with the highest value (FWO: 115 kJ/mol) being that of the blend having the largest microalgae content (coal:microalgae = 3:7).

From an economic and environmental point of view, the possibility of a joint combustion of microalgae biomass and coal in power plants may be interesting, since it allows for the use of existing infrastructures, already equipped with appropriate systems for emissions control and staffed with qualified personnel. In the case of residual microalgae biomass from wastewater treatment, combustion might be the preferred and safest management alternative for its valorization. Other uses, such as fertilizer or animal feed, should only be allowed under strict toxicity controls in order to avoid the incorporation of pollutants, such as trace metals or pharmaceuticals into the soils, ground water, and food chain. Meanwhile, if properly designed and operated, the combustion of microalgae biomass can provide high stability of trace metals in the ashes and complete destruction of organic pollutants.

Overall, the results obtained in this work indicated that, under appropriate conditions, co-combustion of microalgae biomass with coal might be an option to consider. In fact, through the solution of energy and mass balances, it has been recently demonstrated that the smart integration of microalgae culturing with a large scale coal power plant is feasible in terms of net energy ratio (NER) and CO_2 emissions [78]. Still, further work regarding the combustion and co-combustion of microalgae biomass in different types of boilers is actually necessary in view of their practical implementation. In a circular economy context, integrating microalgae culture-wastewater treatment and the thermal valorization of residual biomass is a challenge that would allow for closing the loop for a sustainable microalgae culture.

4. Conclusions

The elemental and proximate analysis, DTG and DSC curves showed remarkable differences between MB and BC, which were mainly related to the relative low fixed-carbon and large volatiles, oxygen and nitrogen content of the first. Despite these differences, both the DTG and DSC combustion curves of the blend MB-BC (wt. 10% MB) were equal to those of BC, which was further corroborated by the corresponding characteristic parameters. The small differences between the experimental and the weighted average composition calculated DTG and DSC curves pointed to unremarkable interactions between MB and BC during the combustion of their blend. On the other hand, the HHV of MB (23 MJ/kg) was very close to that of BC (24 MJ/kg), which is favorable for co-combustion applications. The HHV of MB and BC were adequately estimated by correlations that were based on their elemental analysis or on both their elemental and proximate analysis. Furthermore, the E^* estimated by the iso-conversional models of FWO and KAS for the combustion of MB-BC (103 and 94 kJ/mol, respectively) rather approached those of BC (89 and 79 kJ/mol, respectively) as compared with MB (197 and 196 kJ/mol, respectively). Globally, the results pointed to co-combustion as an encouraging option for the thermal valorization of microalgae biomass resulting from wastewater treatment.

Supplementary Materials: The following are available online at http://www.mdpi.com/1996-1073/12/15/2962/s1.

Author Contributions: Conceptualization and methodology, R.N.C. and M.O.; software and formal analysis, R.N.C.; investigation, R.N.C., C.E. and M.O.; writing—original draft preparation, R.N.C. and M.O.; writing—review and editing, R.N.C., C.E. and M.O.; funding acquisition, M.O.

Funding: This research received funding by Fundação para a Ciência e a Tecnologia (FCT, Lisboa, Portugal), grant number IF/00314/2015.

Acknowledgments: Thanks are due to FCT/MCTES for the financial support to CESAM (UID/AMB/50017/2019), through national funds.

Conflicts of Interest: The authors declare no conflict of interest. The funders had no role in the design of the study; in the collection, analyses, or interpretation of data; in the writing of the manuscript, or in the decision to publish the results.

References

1. *World Energy Resources*; World Energy Council: London, UK, 2016. Available online: https://www.worldenergy.org/wp-content/uploads/2016/10/World-Energy-Resources_SummaryReport_2016.pdf (accessed on 5 July 2019).
2. Assis, T.C.D.; Calijuri, M.L.; Assemany, P.P.; Pereira, A.S.A.D.P.; Martins, M.A. Using atmospheric emissions as CO_2 source in the cultivation of microalgae: Productivity and economic viability. *J. Clean. Prod.* **2019**, *215*, 1160–1169. [CrossRef]
3. Shuba, E.S.; Kifle, D. Microalgae to biofuels: 'Promising' alternative and renewable energy, review. *Renew. Sustain. Energy Rev.* **2018**, *81*, 743–755. [CrossRef]

4. Chen, W.H.; Lin, B.J.; Huang, M.Y.; Chang, J.S. Thermochemical conversion of microalgal biomass into biofuels: A review. *Bioresour. Technol.* **2015**, *184*, 314–327. [CrossRef] [PubMed]
5. Khan, M.I.; Shin, J.H.; Kim, J.D. The promising future of microalgae: Current status, challenges, and optimization of a sustainable and renewable industry for biofuels, feed, and other products. *Microb. Cell Fact.* **2018**, *17*, 36–57. [CrossRef] [PubMed]
6. Saad, M.G.; Dosoky, N.S.; Zoromba, M.S.; Shafik, H.M. Algal Biofuels: Current Status and Key Challenges. *Energies* **2019**, *12*, 1920. [CrossRef]
7. SundarRajan, P.; Gopinath, K.P.; Greetham, D.; Antonysamy, A.J. A review on cleaner production of biofuel feedstock from integrated CO_2 sequestration and wastewater treatment system. *J. Clean. Prod.* **2019**, *210*, 445–458. [CrossRef]
8. Kassim, M.A.; Meng, T.K. Carbon dioxide (CO_2) biofixation by microalgae and its potential for biorefinery and biofuel production. *Sci. Total Environ.* **2017**, *584–585*, 1121–1129. [CrossRef]
9. Delrue, F.; Álvarez-Díaz, P.D.; Fon-Sing, S.; Fleury, G.; Sassi, J.-F. The environmental biorefinery: Using microalgae to remediate wastewater, a win-win paradigm. *Energies* **2016**, *9*, 132. [CrossRef]
10. Razzak, S.A.; Ali, S.A.M.; Hossain, M.M.; deLasa, H. Biological CO_2 fixation with production of microalgae in wastewater—A review. *Renew. Sustain. Energy Rev.* **2017**, *76*, 379–390. [CrossRef]
11. Jais, N.M.; Mohamed, R.M.S.R.; Al-Gheethi, A.A.; Hashim, M.K.A. The dual roles of phycoremediation of wet market wastewater for nutrients and heavy metals removal and microalgae biomass production. *Clean Technol. Environ. Policy* **2017**, *19*, 37–52. [CrossRef]
12. Escapa, C.; Coimbra, R.N.; Paniagua, S.; Garcia, A.I.; Otero, M. Comparison of the culture and harvesting of *Chlorella vulgaris* and *Tetradesmus obliquus* for the removal of pharmaceuticals from water. *J Appl. Phycol.* **2017**, *29*, 1179–1193. [CrossRef]
13. Escapa, C.; Torres, T.; Neuparth, T.; Coimbra, R.N.; García, A.I.; Santos, M.M.; Otero, M. Zebrafish embryo bioassays for a comprehensive evaluation of microalgae efficiency in the removal of diclofenac from water. *Sci. Total Environ.* **2018**, *640–641*, 1024–1033. [CrossRef]
14. Coimbra, R.N.; Escapa, C.; Vázquez, N.C.; Noriega-Hevia, G.; Otero, M. Utilization of non-living microalgae biomass from two different strains for the adsorptive removal of diclofenac from water. *Water* **2018**, *10*, 1401. [CrossRef]
15. Hwang, J.-H.; Church, J.; Lee, S.-J.; Park, J.; Lee, W.H. Use of microalgae for advanced wastewater treatment and sustainable bioenergy generation. *Environ. Eng. Sci.* **2016**, *33*, 882–897. [CrossRef]
16. López-González, D.; Fernandez-Lopez, M.; Valverde, J.L.; Sanchez-Silva, L. Kinetic analysis and thermal characterization of the microalgae combustion process by thermal analysis coupled to mass spectrometry. *Appl. Energy* **2014**, *114*, 227–237. [CrossRef]
17. Lane, D.; Ashman, P.J.; Zevenhoven, M.; Hupa, M.; van Eyk, P.; de Nys, R.; Karlströ, O.; Lewis, D.M. Combustion behavior of algal biomass: carbon release, nitrogen release, and char reactivity. *Energy Fuels* **2014**, *28*, 41–51. [CrossRef]
18. Gai, C.; Liu, Z.; Han, G.; Peng, N.; Fan, A. Combustion behavior and kinetics of low-lipid microalgae via thermogravimetric analysis. *Bioresour. Technol.* **2015**, *181*, 148–154. [CrossRef]
19. Saldarriaga, J.F.; Aguado, R.; Pablos, A.; Amutio, M.; Olazar, M.; Bilbao, J. Fast characterization of biomass fuels by thermogravimetric analysis (TGA). *Fuel* **2015**, *140*, 744–751. [CrossRef]
20. Otero, M.; Calvo, L.F.; Gil, M.V.; García, A.I.; Morán, A. Co-combustion of different sewage sludge and coal: A non-isothermal thermogravimetric kinetic analysis. *Bioresour. Technol.* **2008**, *99*, 6311–6319. [CrossRef]
21. Coimbra, R.N.; Paniagua, S.; Escapa, C.; Calvo, L.F.; Otero, M. Combustion of primary and secondary pulp mill sludge and their respective blends with coal: A thermogravimetric assessment. *Renew. Energy* **2015**, *83*, 1050–1058. [CrossRef]
22. Chen, G.-B.; Chatelier, S.; Lin, H.-T.; Wu, F.-H.; Lin, T.-H. A Study of Sewage Sludge Co-Combustion with Australian Black Coal and Shiitake Substrate. *Energies* **2018**, *11*, 3436. [CrossRef]
23. Coimbra, R.N.; Paniagua, S.; Escapa, C.; Calvo, L.F.; Otero, M. Thermal valorization of pulp mill sludge by co-processing with coal. *Waste Biomass Valoriz.* **2016**, *7*, 995–1006. [CrossRef]
24. Escapa, C.; Coimbra, R.N.; Paniagua, S.; García, A.I.; Otero, M. Comparative assessment of diclofenac removal from water by different microalgae strains. *Algal Res.* **2016**, *18*, 127–134. [CrossRef]

25. Gupta, S.K.; Ansari, F.A.; Shriwastav, A.; Sahoo, N.K.; Rawat, I.; Bux, F. Dual role of *Chlorella sorokiniana* and *Scenedesmus obliquus* for comprehensive wastewater treatment and biomass production for bio-fuels. *J. Clean. Prod.* **2016**, *115*, 255–264. [CrossRef]
26. Mann, J.E.; Myers, J. On pigments growth and photosynthesis of *Phaeodactylum tricornutum*. *J. Phycol.* **1968**, *4*, 349–355. [CrossRef]
27. Organisation for Economic Co-operation and Development (OECD). Test No. 303: Simulation Test—Aerobic Sewage Treatment—A: Activated Sludge Units; B: Biofilms. In *OECD Guidelines for the Testing of Chemicals, Section 3*; OECD Publishing: Paris, France, 2001. [CrossRef]
28. ASTM D3172-13. *Standard Practice for Proximate Analysis of Coal and Coke*; ASTM International: West Conshohocken, PA, USA, 2013.
29. ASTM D3173-11. *Standard Test Method for Moisture in the Analysis Sample of Coal and Coke*; ASTM International: West Conshohocken, PA, USA, 2011.
30. ASTM D3174-12. *Standard Test Method for Ash in the Analysis Sample of Coal and Coke from Coal*; ASTM International: West Conshohocken, PA, USA, 2018.
31. ASTM D3175-18. *Standard Test Method for Volatile Matter in the Analysis Sample of Coal and Coke*; ASTM International: West Conshohocken, PA, USA, 2018.
32. ASTM D5373-08. *Standard Test Methods for Instrumental Determination of Carbon, Hydrogen, and Nitrogen in Laboratory Samples of Coal*; ASTM International: West Conshohocken, PA, USA, 2008.
33. ASTM D4239-14. *Standard Test Method for Sulfur in the Analysis Sample of Coal and Coke Using High-Temperature Tube Furnace Combustion*; ASTM International: West Conshohocken, PA, USA, 2014.
34. UNE-EN 14918. *Solid Biofuels—Determination of Calorific Value*; Spanish Association for Standardization and Certification: Madrid, Spain, 2011.
35. Selvig, W.A.; Gibson, I.H. Calorific value of coal. In *Chemistry of Coal Utilization*; Lowry, H.H., Ed.; Wiley: Hoboken, NJ, USA, 1945; Volume 1, p. 139.
36. Tillman, D.A. *Wood as an Energy Resource*; Academic Press: Cambridge, MA, USA, 1978.
37. Abe, F. The thermochemical study of forest biomass. *Bull. For. For. Prod. Res. Inst.* **1988**, *352*, 1–95.
38. Demirbas, A.; Gullu, D.; Caglar, A.; Akdeniz, F. Estimation of calorific values of fuels from lignocellulosics. *Energy Sources* **1997**, *19*, 765–770. [CrossRef]
39. Sheng, C.; Azevedo, J.L.T. Estimating the higher heating value of biomass fuels from basic analysis data. *Biomass Bioenergy* **2005**, *28*, 499–507. [CrossRef]
40. Yin, C.-Y. Prediction of higher heating values of biomass from proximate and ultimate analyses. *Fuel* **2011**, *90*, 1128–1132. [CrossRef]
41. Jenkins, B.M.; Ebeling, J.M. Thermochemical properties of biomass fuels. *Calif. Agric.* **1985**, *39*, 14–16.
42. Parikh, J.; Channiwala, S.A.; Ghosal, G.K. A correlation for calculating HHV from proximate analysis of solid fuels. *Fuel* **2005**, *84*, 487–494. [CrossRef]
43. Majumder, A.K.; Jain, R.; Banerjee, P.; Barnwal, J.P. Development of a new proximate analysis based correlation to predict calorific value of coal. *Fuel* **2008**, *87*, 3077–3081. [CrossRef]
44. Grabosky, M.; Bain, R. *Properties of biomass relevant to gasification*; Noyes Data Corporation: Park Ridge, NJ, USA, 1981.
45. Bridgwater, A.V.; Double, J.M.; Earp, D.M. Technical and market assessment of biomass gasification in the United Kingdom. In *ETSU Report*; UKAEA: Harwell, UK, 1986.
46. Channiwala, S.A.; Parikh, P.P. A unified correlation for estimating HHV of solid, liquid and gaseous fuels. *Fuel* **2002**, *81*, 1051–1063. [CrossRef]
47. Sajdak, M.; Muzyka, R.; Hrabak, J.; Różycki, G. Biomass, biochar and hard coal: Data mining application to elemental composition and high heating values prediction. *J. Anal. Appl. Pyrol.* **2013**, *104*, 153–160. [CrossRef]
48. Vyazovkin, S.; Wight, C.A. Isothermal and non-isothermal kinetics of thermally stimulated reactions of solids. *Int. Rev. Phys. Chem.* **1998**, *17*, 407–433. [CrossRef]
49. Zhao, M.; Raheem, A.; Memon, Z.M.; Vuppaladadiyam, A.K.; Ji, G. Iso-conversional kinetics of low-lipid micro-algae gasification by air. *J. Clean. Prod.* **2019**, *207*, 618–629. [CrossRef]
50. Ozawa, T. A new method of analyzing thermogravimetric data. *Bull. Chem. Soc. Jpn.* **1965**, *38*, 1881–1886. [CrossRef]

51. Flynn, J.H.; Wall, L.A. A quick, direct method for the determination of activation energy from thermogravimetric data. *Polym. Lett.* **1966**, *4*, 323–328. [CrossRef]
52. Doyle, C.D. Estimating isothermal life from thermogravimetric data. *J. Appl. Polym. Sci.* **1962**, *6*, 639–642. [CrossRef]
53. Kissinger, H.E. Reaction kinetics in differential thermal analysis. *Anal. Chem.* **1957**, *29*, 1702–1706. [CrossRef]
54. Akahira, T.; Sunose, T. Joint convention of four electrical institutes. Research Report (Chiba Institute of Technology). *Sci. Technol.* **1971**, *16*, 22–31.
55. Tsai, M.Y.; Wu, K.T.; Huang, C.C.; Lee, H.T. Co-firing of paper mill sludge and coal in an industrial circulating fluidized bed boiler. *Waste Manag.* **2002**, *22*, 439–442. [CrossRef]
56. Gao, Y.; Tahmasebi, A.; Dou, J.; Yu, J. Combustion characteristics and air pollutant formation during oxy-fuel co-combustion of microalgae and lignite. *Bioresour. Technol.* **2016**, *207*, 276–284. [CrossRef]
57. Chen, C.; Lu, Z.; Ma, X.; Long, J.; Peng, Y.; Hu, L.; Lu, Q. Oxy-fuel combustion characteristics and kinetics of microalgae Chlorella vulgaris by thermogravimetric analysis. *Bioresour. Technol.* **2013**, *144*, 563–571. [CrossRef]
58. Zakariah, N.A.; Rahman, N.A.; Him, N.R.N. Effects of nitrogen supplementation in replete condition on the biomass yield and microalgae properties of *Chlorella sorokiniana*. *ARPN J. Eng. Appl. Sci.* **2017**, *12*, 3290–3298.
59. Paniagua, S.; Calvo, L.F.; Escapa, C.; Coimbra, R.N.; Otero, M.; García, A.I. *Chlorella sorokiniana* thermogravimetric analysis and combustion characteristic indexes estimation. *J. Therm. Anal. Calorim.* **2018**, *131*, 3139–3149. [CrossRef]
60. Babich, I.V.; Hulst, M.V.D.; Lefferts, L.; Moulijn, J.A.; O'Connor, P.; Seshan, K. Catalytic pyrolysis of microalgae to high-quality liquid bio-fuels. *Biomass Bioenergy* **2011**, *35*, 3199–3207. [CrossRef]
61. Xu, L.; Wim Brilman, D.W.F.; Withag, J.A.M.; Brem, G.; Kersten, S. Assessment of a dry and a wet route for the production of biofuels from microalgae: energy balance analysis. *Bioresour. Technol.* **2011**, *102*, 5113–5122. [CrossRef]
62. Wang, K.; Brown, R.C.; Homsy, S.; Martinez, L.; Sidhu, S.S. Fast pyrolysis of microalgae remnants in a fluidized bed reactor for bio-oil and biochar production. *Bioresour. Technol.* **2013**, *127*, 494–499. [CrossRef]
63. Kebelmann, K.; Hornung, A.; Karsten, U.; Griffiths, G. Intermediate pyrolysis and product identification by TGA and Py-GC/MS of green microalgae and their extracted protein and lipid components. *Biomass Bioenergy* **2013**, *49*, 38–48. [CrossRef]
64. Zou, S.; Wu, Y.; Yang, M.; Li, C.; Tong, J. Pyrolysis characteristics and kinetics of the marine microalgae *Dunaliella tertiolecta* using thermogravimetric analyzer. *Bioresour. Technol.* **2010**, *101*, 359–365.
65. Chen, W.H.; Huang, M.Y.; Chang, J.S.; Chen, C.Y. Thermal decomposition dynamics and severity of microalgae residues in torrefaction. *Bioresour. Technol.* **2014**, *169*, 258–264. [CrossRef]
66. Jena, U.; Das, K.C. Comparative evaluation of thermochemical liquefaction and pyrolysis for bio-oil production from microalgae. *Energy Fuels* **2011**, *25*, 5472–5482. [CrossRef]
67. Wu, K.T.; Tsai, C.J.; Chen, C.S.; Chen, H.W. The characteristics of torrefied microalgae. *Appl. Energy* **2012**, *100*, 52–57. [CrossRef]
68. Chen, W.H.; Wu, Z.Y.; Chang, J.S. Isothermal and non-isothermal torrefaction characteristics and kinetics of microalga Scenedesmus obliquus CNW-N. *Bioresour. Technol.* **2014**, *155*, 245–251. [CrossRef]
69. Bui, H.-H.; Tran, K.-Q.; Chen, W.-H. Pyrolysis of microalgae residues—A Kinetic study. *Bioresour. Technol.* **2015**, *199*, 362–366. [CrossRef]
70. Chen, C.; Ma, X.; Liu, K. Thermogravimetric analysis of microalgae combustion under different oxygen supply concentrations. *Appl. Energy* **2011**, *88*, 3189–3196. [CrossRef]
71. Soria-Verdugo, A.; Goos, E.; García-Hernando, N.; Riedel, U. Analyzing the pyrolysis kinetics of several microalgae species by various differential and integral isoconversional kinetic methods and the Distributed Activation Energy Model. *Algal Res.* **2018**, *32*, 11–29. [CrossRef]
72. López, R.; Fernández, C.; Gómez, X.; Martínez, O.; Sánchez, M.E. Thermogravimetric analysis of lignocellulosic and microalgae biomasses and their blends during combustion. *J. Therm. Anal. Calorim.* **2013**, *114*, 295–305. [CrossRef]
73. Chen, C.; Ma, X.; He, Y. Co-pyrolysis characteristics of microalgae *Chlorella vulgaris* and coal through TGA. *Bioresour. Technol.* **2012**, *117*, 264–273. [CrossRef]
74. Chen, C.; Chan, Q.N.; Medwell, P.R.; Heng Yeoh, G. Co-combustion characteristics and kinetics of microalgae *Chlorella vulgaris* and coal through TGA. *Combust. Sci. Technol.*. in press. [CrossRef]

75. Guo, F.H.; Zhong, Z.P. Optimization of the co-combustion of coal and composite biomass pellets. *J. Clean. Prod.* **2018**, *185*, 399–407. [CrossRef]
76. Ondro, T.; Vitázek, I.; Húlan, T.; Lawson, M.K.; Csáki, Š. Non-isothermal kinetic analysis of the thermal decomposition of spruce wood in air atmosphere. *Res. Agric. Eng.* **2018**, *64*, 41–46.
77. Zhao, Z.; Liu, P.; Wang, S.; Ma, S.; Cao, J. Combustion characteristics and kinetics of five tropic algal strains using thermogravimetric analysis. *J. Therm. Anal. Calorim.* **2018**, *131*, 1919–1931. [CrossRef]
78. Giostri, A.; Binotti, M.; Macchi, E. Microalgae cofiring in coal power plants: Innovative system layout and energy analysis. *Renew. Energy* **2016**, *95*, 449–464. [CrossRef]

energies

MDPI

Article

Optimization of Air Distribution to Reduce NOx Emission and Unburned Carbon for the Retrofit of a 500 MWe Tangential-Firing Coal Boiler

Hyunbin Jo [1], Kiseop Kang [1], Jongkeun Park [1], Changkook Ryu [1,*], Hyunsoo Ahn [2] and Younggun Go [2]

1 School of Mechanical Engineering, Sungkyunkwan University, Suwon 16419, Korea
2 Doosan Heavy Industries and Construction, Youngin 16858, Korea
* Correspondence: cryu@me.skku.ac.kr; Tel.: +82-31-290-4841

Received: 18 July 2019; Accepted: 23 August 2019; Published: 26 August 2019

Abstract: The use of separated overfire air (SOFA) has become a standard technique of air staging for NOx reduction in the coal-fired boiler and can also be applied to existing boilers by retrofit. This study was to optimize the air distribution for the proposed SOFA installation in a 500 MWe tangential-firing boiler that has 20 identical units in Korea. Using computational fluid dynamics (CFD) incorporating advanced coal combustion submodels, the reference case was established in good agreement with the design data, and different flow ratios of burner secondary air, close-coupled OFA (CCOFA), and SOFA were evaluated. Increasing the total OFA ratio effectively suppressed NO formation within the burner zone but had a negative impact on the boiler performance. With moderate air staging, NO reduction became active between the CCOFA and SOFA levels and, therefore, the OFA distribution could be optimized for the overall boiler performance. For total OFA ratios of 25% and 30% with respective burner zone stoichiometric ratios of 0.847 and 0.791, increasing the SOFA ratio to 15% and 20%, respectively, was ideal for decreasing the unburned carbon release and ash slagging as well as NO emission. Too high or low SOFA ratios rapidly increased the unburned carbon because of inefficient mixing between the strong air jets and char particles. Based on these ideal cases, the actual air distribution can be adjusted depending on the coal properties such as the ash slagging propensity.

Keywords: coal; combustion; computational fluid dynamics; boiler; overfire air; NOx emission

1. Introduction

NOx emission from coal-fired power plants has become a crucial issue in the power industry in Korea because of its contribution to the formation of secondary particulate matter of less than 2.5 μm (PM2.5) by photochemical reactions. Due to climate change in the Korean Peninsula and the increase of inbound pollutants, severe haze events have become frequent in recent years [1,2]. The contribution of secondary PM originating from coal-fired power plants is yet to be clarified among the inbound and domestic sources. However, air quality concern has a negative impact on the public perception of coal power and has changed energy policy including the temporary shutdown of old plants, as well as more stringent emission regulations. Although advanced combustion technology and efficient gas cleaning equipment are already in use, further lowering the pollutant emission from existing power plants to the minimum has become a most urgent issue. This applies to existing power plants, which include 20 units of 500 MWe tangential-firing (TF) boilers in Korea that have been built since 1993 with an identical design and are referred to as the 'standard' 500 MWe unit.

NOx emission reduction has been a key topic of combustion and gas cleaning technology. As the primary measure, the formation of NOx needs to be minimized by a combination of air staging, fuel staging, and low-NOx burners [3]. NOx can then be removed by secondary measures such as selective catalytic or non-catalytic reduction. In particular, NOx emission from coal is dominated by the fuel NOx mechanism, and therefore optimizing the reaction stoichiometry by air staging is very effective for its reduction. One advanced technique of air staging is the use of separated overfire air (SOFA) injected distantly above the burner zone. It is differentiated from close-coupled OFA (CCOFA) which is injected immediately above the top burners in a TF boiler. Adjusting the CCOFA ratio can be helpful in reducing NOx [4], but multi-level air staging by the use of SOFA can provide increased retention time for the reduction reactions to be more effective. The SOFA technique can also be applied to the 20 standard TF boilers in Korea by retrofit.

Together with NO emission, furnace exit gas temperature (FEGT) and unburned carbon (UBC) in ash are the key performance parameters of boiler operation associated with the operability and efficiency [5]. Various studies of the optimization of SOFA to improve boiler performance have been reported in the literature based on experiments and/or computational fluid dynamics (CFD). Increasing the SOFA ratio and optimizing its detailed distribution were found to be effective in reducing NOx emission for various boiler types [6–13]. However, the change in reaction stoichiometry can also significantly alter the combustion and heat transfer characteristics. In the study of Zha et al. [9] in a 600 MWe TF boiler, increasing the SOFA ratio from 10% to 40% achieved 50% reduction in NOx emission with more uniform heat flux distribution. However, this accompanied an unfavorable increase in the UBC in the fly ash, in the CO concentration, and in the flue gas temperature at the platen superheater, together with a significant change in the heat absorption pattern between the wall and convective heat exchangers. Li et al. [14] performed experiments on various damper openings of secondary air (SA) in the burner zone, CCOFA, and SOFA for a retrofitted 300 MWe TF boiler, and found that an ideal setup reduced the NOx emission by 44% with the UBC in ash not influenced at a sacrifice in boiler efficiency of 0.21%. The aerodynamics of SOFA also has a large influence on the UBC and heat absorption pattern in the heat exchangers downstream and adjusting the yaw and tilt angles of SOFA can alleviate such problems [15–17]. The reducing atmosphere in the burner zone by increasing the SOFA ratio may increase the possibility of fireside corrosion by H_2S. In this respect, the experimental study of Xu et al. [18] reported that the air distribution can be adjusted to reach a balance between low corrosion, low NOx emission, and high boiler efficiency.

This study was to optimize the flow rate distribution of CCOFA and SOFA for NOx reduction in the standard 500 MWe TF boiler to be retrofitted including the installation of SOFA. Different ratios of flow rates between the burner secondary air, CCOFA, and SOFA were simulated. The CFD method was validated using the design data for the reference case of the retrofit boiler. From the results, the boiler performance was evaluated to determine the ideal flow ratios, and to understand the reasons for the differences in terms of NOx emission, UBC in ash, furnace exit gas temperature (FEGT), heat transfer rates, boiler efficiency, and the possibility of high-temperature corrosion.

2. Target Boilers and Numerical Methods

2.1. Target Boiler and Operation Conditions

Figure 1 shows a schematic of the 500 MWe coal-fired TF boiler modified from its original design. The modification was intended to adjust the operation range to low-rank coals and to improve its efficiency with an increased steam temperature from 538 °C to 596 °C. Because the original boiler only had CCOFA for air staging, SOFA was to be installed for the efficient reduction of NOx emission. This retrofit was also expected to extend its lifetime by ten years. The burner zone had six levels of coal burners (A to F levels) installed on the corners to create a swirling flow (fireball) at the center. It was divided into three sections with two burners each having an identical arrangement of coal and air supply ports as illustrated in the figure. Each coal burner aerodynamically split the coal and primary air flow into concentrated and weak ports depending on the coal concentration. The CCOFA was injected through four ports immediately above the top burner F on each corner. In the retrofit design, the SOFA was located 6.144 m above the CCOFA with two ports on each corner and one each on the adjacent side walls. The convective heat exchangers in the upper part had a new arrangement consisting of superheaters (SH) and reheaters (RH) with different tube bank geometry and steam conditions. An economizer (ECO) was located in the backpass. The furnace was built with membrane tube walls that act as an evaporator of the preheated water from the ECO. The new steam pressure and temperature leaving the final SH to the high-pressure turbine were 255.4 kg/cm^{2}g and 596 °C, respectively.

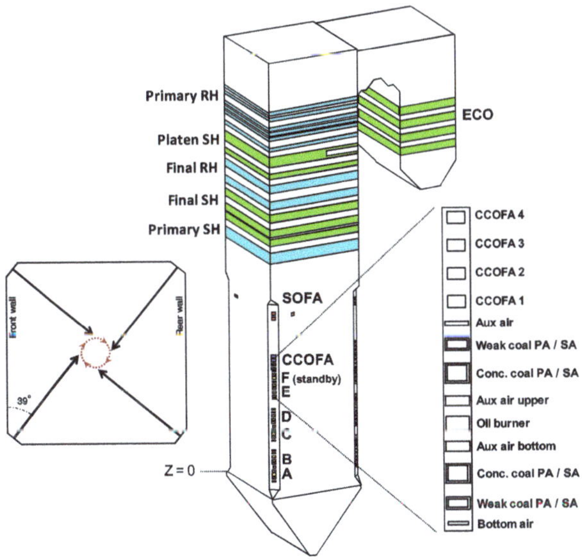

Figure 1. Schematic of the 500 MWe coal-fired tangential-firing (TF) boiler.

Table 1 presents the characteristics of the performance coal and summary of the operating conditions. The coal for the retrofit boiler was sub-bituminous with a higher heating value (HHV) of 5600 kcal/kg which was significantly lower than that for the old design coal (6300 kcal/kg). At the nominal rate load, the coal throughput was 55.583 kg/s which delivered 1303 MW$_{th}$ of thermal input on an HHV basis. The coal was pulverized to an average particle size of 50 μm, and partially dried by hot primary air to have a moisture content of 8.6% and a mass flow rate of 50.494 kg/s. It was transported by the primary air (including the evaporated moisture) from the pulverizers to the burners a to E. The secondary air (SA) with a total flow rate of 355.1 kg/s at 326 °C was distributed into different ports of the burner SA, CCOFA, and SOFA. The overall excess air ratio was 12.94%.

Table 1. Operating conditions of the 500 MWe coal-fired boiler.

Input	Values
Coal properties	Proximate analysis (% wet): Total moisture 17, volatile matter 31.36, fixed carbon 43.89, ash 7.75 Ultimate analysis (% daf): C 76.95, H 5.34, O 16.01, N 1.19, S 0.51 Higher heating value (HHV): 5600 kcal/kg
Coal throughput	55.583 kg/s (50.494 kg/s after drying/pulverization)
Primary air	121.2 kg/s, 224 °C (126.3 kg/s, 77 °C after drying/pulverization)
Total secondary air	355.1 kg/s, 326 °C
Excess air ratio	12.94%

Table 2 summarizes the simulation cases for various distributions of CCOFA and SOFA. In the first set of cases, the SOFA ratio was increased from 15% to 35%, while fixing the CCOFA ratio at

5%. These values corresponded to the burner zone stoichiometric ratio (SR) of 0.904 to 0.678. In the second set, the distribution between CCOFA and SOFA was varied from 0% to 25% by 5% increments, while fixing the total OFA ratio at 25%. Each case was named after the ratio in the two OFA ports. For example, C05-S20 refers to the case with 5% CCOFA ratio and 20% SOFA ratio. This is the reference case to be used for validation of the CFD results by comparison with the design data.

Table 2. Cases of various close-coupled overfire air (CCOFA) and separated overfire air (SOFA) ratios for computational fluid dynamics (CFD) simulations.

Case Set	Case Name	Ratio in Total Combustion Air (%)			Burner Zone SR	Note
		CCOFA	SOFA	Total OFA		
Set #1	C05-S15	5	15	20	0.904	
	C05-S20		20	25	0.847	Reference case
	C05-S25		25	30	0.791	
	C05-S30		30	35	0.734	
	C05-S35		35	40	0.678	
Set #2	C25-S00	25	0	25	0.847	
	C20-S05	20	5			
	C15-S10	15	10			
	C10-S15	10	15			
	C05-S20	5	20			Reference case
	C00-S25	0	25			
Set #3	C30-S00	30	0	30	0.791	
	C25-S05	25	5			
	C20-S10	20	10			
	C15-S15	15	15			
	C10-S20	10	20			
	C05-S25	5	25			
	C00-S30	0	25			

2.2. CFD Modeling

The mesh was constructed for the boiler using 3,529,358 cells. This mesh was selected after assessing the sensitivity by comparing the degree of numerical diffusion between coarser (1.23 million cells) and finer (5.47 million cells) meshes. In these meshes, the cell fineness was varied only in the burner zone to have a respective average volume of 0.0114, 0.0034, and 0.0022 m^3/cell. Compared to the finer mesh, the selected one exhibited very small deviations in the key performance parameters (0.12% in carbon conversion, 1.9% in the exit NO concentration, and 0.2% in the heat absorption on the furnace wall). In the detailed comparison for the flow pattern along the vertical centerline, the deviations from the results of the finer mesh were 4.7% in the velocity profile and 0.71% in the temperature profile. By contrast, the coarser mesh had deviations of 17.58% and 1.11%, respectively. The computation time for the selected mesh to reach a converged solution was 69% of that for the finer mesh. The details of the mesh sensitivity test together with the iteration strategy have been reported elsewhere [19].

The CFD simulations were performed using ANSYS Fluent (version 17.2) [20] with submodels for reactions, turbulence, and radiation, most of which have been reviewed by Sankar et al. [5] as common models applied to a coal-fired boiler. The reaction submodels are summarized in Table 3. Coal particles were tracked using the discrete phase method for 10 size fractions ranging between 5.9 μm and 204 μm, with a total of 49,560 particles. FLASHCHAIN [21] was used to determine the input parameters for coal devolatilization in a drop tube furnace at 1200 °C which was close to the heat transfer condition of pulverized particles in the boiler. This code predicted the reaction dynamics based on the semi-empirical coal network mode to provide the product yields (tar, CO, CO_2, H_2O, CH_4, C_2H_4, C_2H_6, C_3H_6, HCN, H_2S, and solid char) and reaction kinetics as listed in the table. Because the heating rate was much faster than that in the proximate analysis, the total volatiles yield (58.56% daf) was larger than the volatile matter content (41.67% daf) in Table 1. Because of sharing the same mass source from devolatilization and having similar reaction rates, C_{2+} hydrocarbons were simplified to an imaginary species of C_xH_y. The yield and composition of tar were modified to include the H and O fractions of char so that the char could be modelled as a pure carbon solid and ash. For simplicity, it also incorporated the two minor volatiles (HCN and H_2S). Char conversion by O_2, CO_2, and H_2O was solved using the unreacted core shrinking model (UCSM) [22] which is suitable for high-temperature reactions in the industrial-scale furnace and the release of the UBC in fly ash. This model considers the three competing rates of the heterogeneous reaction on the char core surface and the gas diffusions onto the particle and through the ash layer. With the UCSM, the decrease in the reactivity toward the end of char conversion can be simulated, which was required to predict the UBC in fly ash. The multiple volatile products and the UCSM were implemented into the CFD code using user defined functions (UDF).

Table 3. Summary of submodels adopted for boiler simulation.

Category	Submodels
Coal combustion	-Devolatilization: FLASHCHAIN [21] Dry coal→58.56 wt% daf Volatiles + 32.91 wt% daf C(s)(Char) Composition of Volatiles: Tar 32.57, CO 3.31, CO_2 3.89, H_2O 14.35, H_2 0.71, CH_4 1.99, C_xH_y 1.73 wt% daf Devolatilization rate: $\frac{dV}{dt} = A\ \exp\left(-\frac{E}{RT}\right)(V_0 - V)$; E = 5.49 kcal/mol, a = 7.89 × 103 s^{-1} -Char combustion: Unreacted core shrinking model [22] $R_{char,i} = \frac{1}{\frac{1}{k_{diff,i}} + \frac{1}{k_{s,i}Y^2} + \frac{1}{k_{dash,i}}\left(\frac{1}{Y} - 1\right)}\left(P_i - P_i^*\right)\left[\text{g cm}^{-2}\ \text{s}^{-1}\right]$ $k_{dash} = k_{diff}\varepsilon^{2.5}, Y = d_{char}/d_p$ (R1) C(s) + 0.5 O_2 → CO $k_s = 8710 \exp(-17967/T_s),\ k_{diff} = 1.383 \times 10^{-3}\left(\frac{T}{1800}\right)^{0.75}/\left(P_t d_p\right)$ $P_i - P_i^* = P_{O_2}$ (R2) C(s) + H_2O → CO+H_2 $k_s = 247 \exp(-21060/T_s),\ k_{diff} = 1 \times 10^{-3}\left(\frac{T}{2000}\right)^{0.75}/\left(P_t d_p\right)$ $P_i - P_i^* = P_{H_2O} - (P_{H_2} \cdot P_{CO})/K_{eq},\ K_{eq} = \exp[17.644 - 30260/(1.8T_s)]$ (R3) C(s) + CO_2 → 2 CO $k_s = 247 \exp(-21060/T_s),\ k_{diff} = 7.45 \times 10^{-4}\left(\frac{T}{2000}\right)^{0.75}/\left(P_t d_p\right)$ $P_i - P_i^* = P_{CO_2}$
Species, gas reaction	-Species: Tar, CO, CO_2, H_2, CH_4, C_xH_y, H_2, SO_2, O_2, N_2 -Reaction mechanism [23,24] (R4) $C_xH_yO_z$ (tar) + a O_2 → x CO + 0.5y H_2 (R5) C_nH_m + 0.5n O_2 → n CO + 0.5m H_2 (R6) C_nH_m + 0.5n H_2O → n CO + 0.5(m+n) H_2 (R7) CH_4 + 0.5 O_2 → CO + 2 H_2 (R8) CH_4 + 0.5 H_2O → CO + 2.5 H_2 (R9) CO + H_2O → CO_2 + H_2 (R10)H_2 + 0.5 O_2 → H_2O -Reaction rate: kinetic rate/eddy dissipation rate model [25]
Discrete phase	-Lagrangian scheme with stochastic tracking for turbulence -Number of particles: 49,560 -Particle size: 5.88–204 μm
NOx	-Thermal NOx: Extended Zeldovich mechanism [26] -Fuel NOx: De Soete [27] -Fuel-N intermediate: HCN 83.33%, NH_3 16.67%

Gaseous reactions (R4)–(R10) were based on the global reaction scheme of Jones and Lindstedt [23] for hydrocarbon and the tar oxidation rate for (R4) was taken from Smoot and Smith [24]. In the reaction rate, the turbulence-chemistry interaction was considered using the kinetic rate/eddy dissipation rate model [25] with the realizable k-ε model employed for turbulence.

Regarding the heat transfer, radiation was solved using the discrete ordinate method with the weighted sum of the gray gases model for gas absorption [28]. In the boundary condition, the entire furnace wall (evaporator) was set to have the average water/steam temperature of 652.15 K with an overall heat transfer coefficient of 250 W/m^2K and inner wall emissivity of 0.7. The heat transfer coefficient was taken as the average of measured values 3.51–4.37 m^2K/kW of thermal resistance equivalent to 229–285 W/m^2K depending on the coal types [29]. The tube bundles from primary SH to the economizer were simplified as porous zones, with local source terms calculated for flow resistance, convection, and radiation based on the detailed geometry and steam conditions [30]. In brief, the inertial resistance in the lateral and transverse directions was calculated using the Jakob correlation [31]:

$$\Delta \mathrm{p} = \frac{2f' {G_{max}}^2 N_L}{\rho}\left(\frac{\mu_s}{\mu}\right)^{0.14} \tag{1}$$

$$f' = [0.044 + \frac{0.08\left(\frac{S_L}{D}\right)}{\left(\frac{S_T}{D} - 1\right)^{0.43+1.13D/S_L}}]Re^{-0.15} \tag{2}$$

Similarly, the heat transfer rate in each cell was calculated as the sum of convection ($\dot{q}_{conv}$) and radiation ($\dot{q}_{rad}$) for the specific tube surface area per unit volume (A_s):

$$\dot{q}_{conv} = f_{conv} A_s \frac{k\,Nu}{D}\left(T_{gas} - T_s\right)\left[\mathrm{W/m^3}\right] \tag{3}$$

$$\dot{q}_{rad} = f_{rad}\varepsilon\sigma A_s\left({T_{gas}}^4 - {T_s}^4\right)\left[\mathrm{W/m^3}\right] \tag{4}$$

The Nusselt number (Nu) for convection was determined from the Zukauskas correlation [32]:

$$Nu = 0.40{Re_D}^{0.6}Pr^{0.36} \tag{5}$$

The above equations include two correction factors (f_{conv} and f_{rad}) that were introduced to consider the slagging/fouling factors and tuned to match the design values. The formulations for the flow resistance and heat transfer rates were implemented using UDFs.

NO reactions were calculated by post-processing of the CFD results. Thermal NO reactions were based on the extended Zeldovich mechanism with the rate constants taken from Hanson and Salimian [26] and the radical concentrations acquired from the respective partial equilibrium assumptions. In the fuel NO mechanism [27], the partitioning of fuel-N between volatile-N and char-N was determined using FLASHCHAIN. The release of N intermediates during the devolatilization of the low rank coal was assumed to be HCN 5:NH_3 1, whereas the char-N was converted directly to NO. The reduction of NO to N_2 on the active char surface was considered with a microscopic surface area that was assumed to be 100 m^2/g. Prompt NO was ignored because its concentration is known to be minor [3].

3. Results

3.1. Reaction and Heat Transfer in the Reference Case

The key results for the reference case (C05-S20) are explained in this section to understand the flow, reaction, and heat transfer characteristics of the boiler before comparisons between the simulation cases varying the CCOFA and SOFA ratios. As shown in Figure 2, the flow and combustion patterns were characterized by the formation of a large swirling fireball across the burner zone which is typical for a TF boiler. The fireball was pushed to the center by the large momentum delivered by the jets from the burners on the corners, whereas it expanded towards the wall by the centrifugal force in the spacings between the burners B and C, and D and E. This created the crescent shapes of a high velocity region above 20 m/s in the vertical cross-section of Figure 2a. With the increase of flow rate from the burners a to E, the fireball size became larger. Above the SOFA, the fireball finally expanded, before the flow entered the tube bundles. The path-lines from the burner a in Figure 2b show that the flow from this burner filled the bottom section, and then moved upward through the center of the fireball. The jet flows from the other burners above joined the swirling flow of the fireball mostly at its outer part.

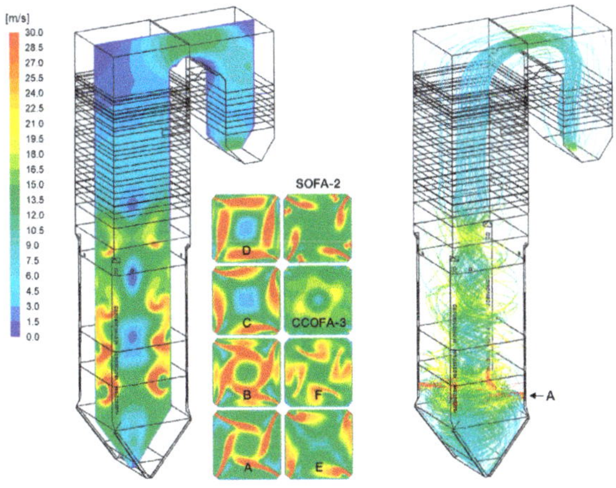

(a) (b)

Figure 2. Velocity contours and path-line from burner a for the reference case (C05-S20) (**a**) Velocity and (**b**) Path-line.

Figure 3a shows the temperature on the selected cross-sections and heat flux on the wall. The bottom cone region was filled with the flow from burner a and was cooled down to a low temperature by heat transfer to the wall. It then entered the center of the fireball, which created the bell-shaped low-temperature (<1000 °C) region in the lower furnace. The temperature gradually increased by introduction of more coal along the height to form the high-temperature zone above 1500 °C, which stretched from the near-wall region of the burner D level to the center at the SOFA level. The cold SOFA jets at 326 °C caused a temperature drop by dilution although it delivered additional air for burnout of the remaining char and combustible gases. The temperature decreased rapidly above the primary SH which virtually stopped further gaseous or heterogeneous reactions. The wall heat flux shown in Figure 3b was large in the burner zone with a peak of 220.6 W/m^2 appearing on the burner D level which coincided with the temperature contours.

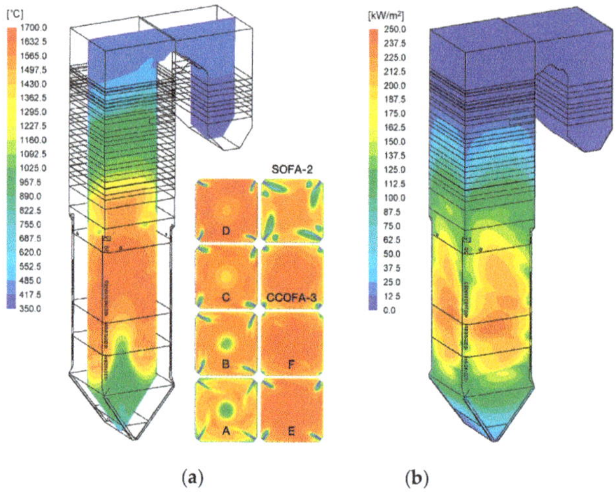

Figure 3. Temperature and wall heat flux for the reference case (C05-S20). (**a**) Temperature and (**b**) Wall heat flux.

Figure 4 shows the solid carbon concentration and the rates of char conversion reactions. The solid carbon in char particles had high concentrations (>0.01 kg/m^3) close to the wall caused by the inertia of the swirling flow. In addition, some char particles originating from the lowermost burners filled the bottom cone and then entered the center of the fireball from below with insufficient char conversion. Because the burner zone was fuel-rich (SR of 0.847), the solid carbon remained at the top of the burner zone. The CCOFA and SOFA delivered the rest of the air for char oxidation. Although the overall SR became fuel-lean, solid carbon was present in the heat exchanger zone because the mixing by the OFA jets was not perfect. In particular, unburned char particles were present mostly on the corners of cross-section (1) with a concentration over 5×10^{-5} kg/m^3 in Figure 4a. On cross-section (2), the carbon concentration decreased along the SOFA paths to below 2×10^{-6} kg/m^3 by slow oxidation but the corners still had concentrated char particles. Above this section, the temperature was not high enough for further char reactions, which led to the UBC in fly ash. The contour on the vertical cross-section in Figure 4a appears to have an increased carbon concentration in the top furnace because the mixing of the particles from the corners with the gas flow slowly progressed. Therefore, the trajectories of OFA jets and its mixing with the char particles are important in reducing the UBC before they enter the heat exchangers.

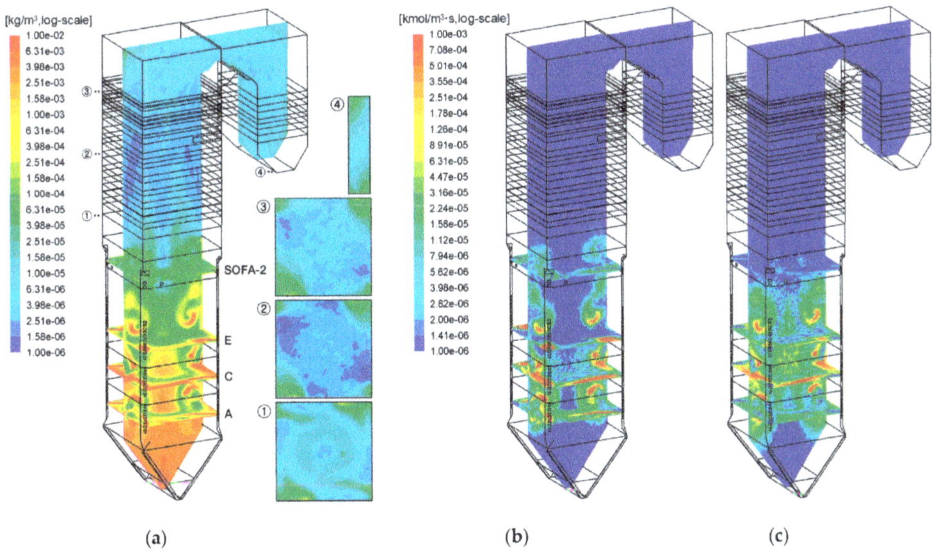

Figure 4. Solid carbon concentration and char conversion rates for the reference (C05-S20). (**a**) Solid carbon concentration; (**b**) Char-O_2 rate and (**c**) Char-CO_2 rate.

As shown in Figure 4b, the char oxidation reactions were active in the outer part of the fireball where most char particles were present, and the combustion air was delivered. However, no oxidation occurred in the bottom cone, at the center of the fireball, and above the primary SH, because the temperature was not high enough and/or O_2 was depleted. Because the temperature was high enough in the burner zone, the gasification by CO_2 and H_2O also made a significant contribution to the char conversion (Figure 4c). Integrating the reaction rates over the entire volume, the two gasification reactions accounted for 36.7% of char conversion and the remainder by oxidation.

Figure 5a shows that O_2 delivered by the burner SA was consumed rapidly and was depleted in most of the burner zone. By contrast, O_2 from the SOFA penetrated deeper, because in this case, the SOFA ratio was relatively large (20%). The excess O_2 then approached the wall as the fireball expanded. O_2 was fully consumed at the central region above the SOFA to have a very low solid carbon concentration. After the flow entered the heat exchanger zones, the fireball quickly disappeared, and the mixing slowly progressed. The CO mole fraction above 5% appeared along the path of coal particles in the burner zone, as shown in Figure 5b. In particular, its concentration was very high on the sidewalls of burners C and D. The excessively reducing atmosphere can increase the possibility of high-temperature corrosion of the wall. Without further fuel supply, the CO mole fraction decreased rapidly by the CCOFA injection and became below 0.25% on the SOFA level. On the platen SH, the CO concentration was 46 ppm.

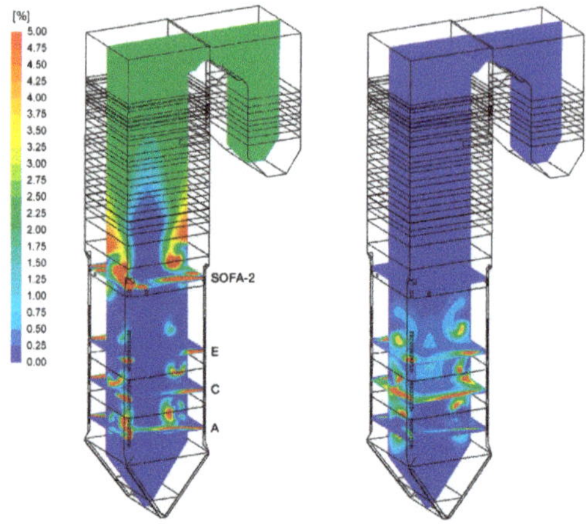

Figure 5. Mole fractions of O_2 and CO for the reference case (C05-S20). (**a**) O_2 mole fraction and (**b**) CO mole fraction.

Figure 6 shows the reaction rate and concentration of NO. Along the trajectories of coal particles, the fast devolatilization with very high heating rate released the N-intermediates which were partially oxidized to produce NO under the oxygen-rich atmosphere with the PA and burner SA around. This was followed by char conversion releasing NO directly from char-N. This led to the regions of fast NO formation rates of over 1×10^{-6} kmol/m^3·s, which coincided with those of high concentrations of solid carbon (Figure 4a) and O_2 (Figure 5a). Outside these regions, however, O_2 was depleted and the reducing atmosphere caused the removal of NO (negative reaction rates). In particular, rapid reduction reactions ($<-1 \times 10^{-6}$ kmol/m^3·s) took place by the remaining N-intermediates and by char with the already produced NO in the region around the coal jets where O_2 was depleted. Also, the volume between the CCOFA and SOFA provided the spaces for further reduction reactions of NO, mainly by residual char. Integrating the reaction rates, the NO emission was dominated by the fuel NO mechanism, and the contribution of the thermal NO mechanism was only approximately 10%.

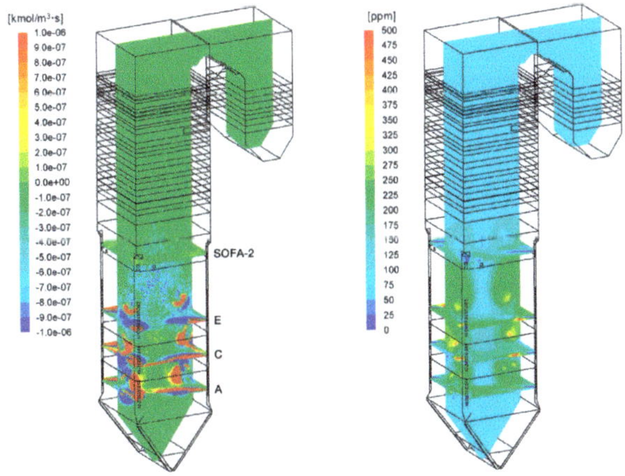

(a) **(b)**

Figure 6. Reaction rate and concentration of NO for the reference case (C05-S20). (**a**) NO rate and (**b**) NO concentration.

Table 4 compares the design data and CFD results for the reference case. The difference of O_2 concentration between by design and the CFD result was 0.02%, owing to the release of UBC in fly ash and bottom ash. The predicted NO concentration was below the guaranteed value. The predicted heat absorption on the furnace wall using the measured thermal resistance was very close to the design value with a deviation of 0.04 MW_{th}. The heat absorption on the tube bundles calibrated using f_{conv} and f_{rad} was reasonably close to the design values. These results also imply that the predicted temperature distribution along the furnace height would be reasonable. However, this study was for the proposed boiler retrofit and the modeling approach was not validated by experiments. Therefore, the focus was on the comparative evaluation of the key performance parameters between simulation cases.

Table 4. Comparison of design data and CFD results for the reference case.

Parameter		Design	CFD
Exit O_2 (% dry)		2.45	2.47
Exit NO (ppm, 6% O_2)		<150	99.75
Heat absorption (MW_{th})	Furnace wall	562.31	562.27
	Primary SH	88.74	84.00
	Platen SH	102.81	99.96
	Final SH	113.04	108.75
	Primary RH	128.86	129.59
	Final RH	75.36	71.05
	Economizer	72.92	73.29

3.2. *Influence of SOFA Ratios*

Figure 7 compares the mass-weighted average profiles of temperature, O_2, solid carbon, and NO concentrations for various SOFA ratios at a fixed CCOFA ratio of 5% and total OFA ratio of 25% (case set #1 in Table 1). In the burner zone, the temperature and O_2 profiles were spread with a deviation of approximately 150 °C and 1%, respectively, with an increase in the SOFA ratio and corresponding decrease in the burner zone SR. The trends were inverted at the SOFA level by fast oxidation reactions of combustible gas species. The temperature profile influenced the distribution of heat absorption between the wall and tube bundles, which will be presented later. The carbon concentration exhibited acute changes responding to the fuel air supply from each coal burner in the burner zone. On the burner E, it varied from 0.00095 kg/m^3 (C05-S15) to 0.00175 kg/m^3 (C05-S35). With no further fuel supply above this level, it showed a continuous decrease to a value below 0.0001 kg/m^3, with a final carbon conversion of 99.92% (C05-S15) and 99.80% (C05-S35). However, the differences between the cases in terms of UBC in fly ash were significant, which will be presented later.

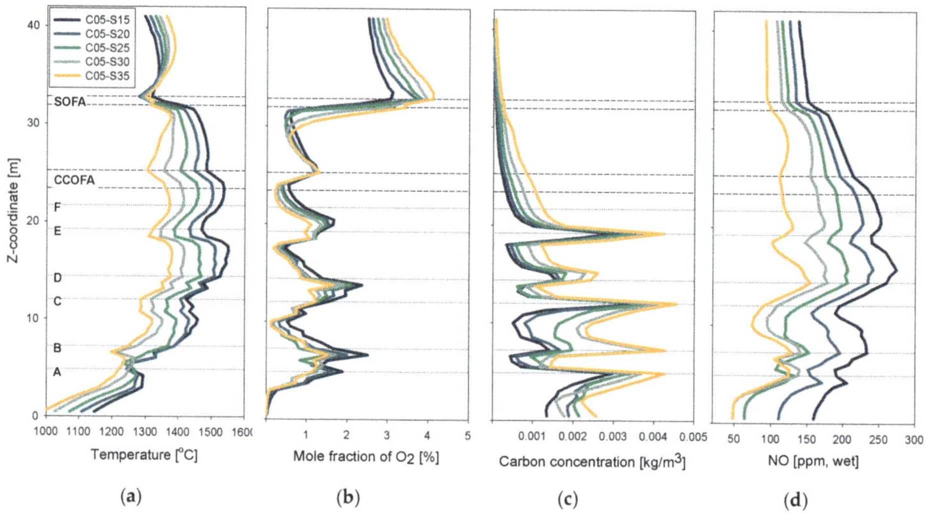

Figure 7. Mass-weighted average profiles of gas temperature and concentrations of O_2, solid carbon, and NO in the burner zone for different SOFA ratios (CCOFA ratio fixed at 5%). (**a**) Temperature; (**b**) O_2; (**c**) Solid carbon and (**d**) NO.

In Figure 7d, the NO concentration sharply increased between the burners A–E, owing to the dominant contribution of fuel NO. Comparing the value between cases, both the thermal and fuel NO formations were suppressed at higher SOFA ratios having lower O_2 concentration and temperature. In particular, NO reduction reactions were active in the region where O_2 was locally depleted, as indicated by the decreases in the concentration between the burner levels of B and C, and D and E. On the burner F (standby) level where all the fuel has already been introduced, NO concentration was 128 ppm for C05-S35 and 251 ppm for C05-S15. However, between the volume between the F burner and the first SOFA ports, low SOFA ratios exhibited active NO reduction reactions which lowered the NO concentration to 181 ppm (−70 ppm from the value on the F level) for C05-S15. By contrast, NO concentration of C05-S35 on the first SOFA level was 110 ppm (−18 ppm). Then, the SOFA injection through the two port levels had an immediate dilution effect that further lowered the NO concentration to 151 ppm (−30 ppm) for C05-S15 and 93 ppm (−17 ppm) for C05-S35. Above the SOFA level, NO concentration was not significant. These results suggest that a high level of air staging is effective for low NO concentration within the burner zone and, therefore, the OFA distribution between the CCOFA and SOFA would not be crucial. However, for a moderate level of air staging, securing the volume (i.e., time) for NO reduction reactions is essential by the installation of SOFA and increasing its ratio larger than the CCOFA ratio.

Although large SOFA ratios effectively reduced NOx emission, negative impacts were accompanied on the boiler performance. Figure 8 shows the NOx emission and UBC in fly ash at the boiler exit and FEGT. FEGT was determined from the average gas temperature entering the first tube bundle (primary SH). The UBC represents the boiler efficiency, while the FEGT is associated with the propensity of ash slagging on the tube bundles. NOx emission was reduced from 109.9 ppm to 73.2 ppm (a 33.4% reduction) on a 6% O_2 basis by the increase in the SOFA ratio, but the FEGT was increased by 62.6 °C, while the UBC was more than tripled. Therefore, the use of SOFA ratio as high as 35% was not favorable.

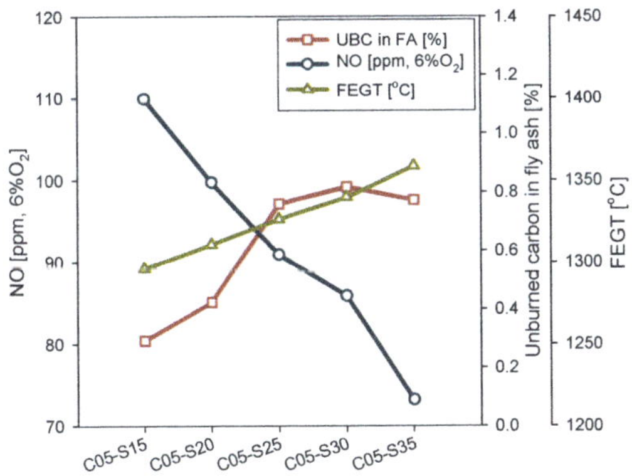

Figure 8. Comparison of NOx emission, unburned carbon (UBC) in fly ash, and furnace exit gas temperature (FEGT) for different SOFA ratios (CCOFA ratio fixed at 5%).

Figure 9 compares the heat absorption by heat exchangers for different SOFA ratios, which was closely associated with the temperature profile shown in Figure 7. On increasing the SOFA ratio, the heat absorption by the furnace wall was decreased from 599 MW_{th} for C05-S15 to 524 MW_{th} for C05-S35 by the lower gas temperatures of the burner zone. Because the trend in the temperature was inverted above the SOFA level, the heat absorption in the tube bundles of SHs, RHs, and ECO increased from a total of 543 MW_{th} for C05-S15 to 614 MW_{th} for C05-S35. The resultant boiler efficiency was 87.61%–87.30% on an HHV basis, which is also plotted in Figure 9. Together with the lower boiler efficiency, the increase in the SHs and RHs at the larger SOFA ratios has an unfavorable impact on the boiler operation. This is because it increases the possibility of high steam temperature at the final SH and RH exits that requires more water spray by the attemperator to maintain the value below the limit.

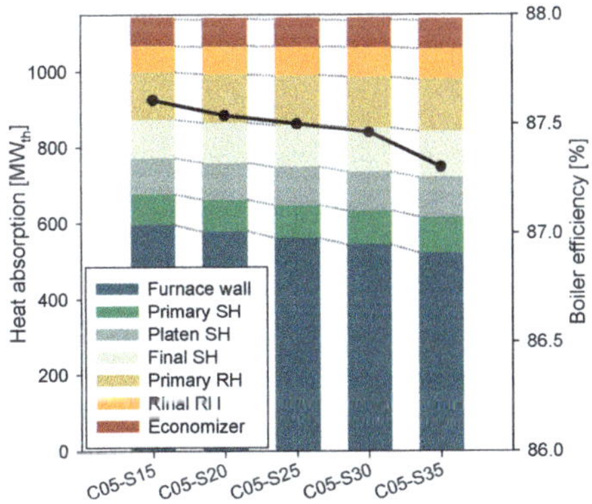

Figure 9. Comparison of heat absorption by heat exchangers and boiler efficiency for different SOFA ratios (CCOFA ratio fixed at 5%).

One additional issue to assess is the influence on corrosion of the water wall. With the increase in the OFA ratio, the burner zone SR decreased from 0.90 (C05-S15) to 0.68 (C05-S35). The more reducing environment in the furnace increases the possibility of high-temperature corrosion by the presence of H_2S, COS, and CO [33], which shortens the lifetime of the boiler. As summarized in Table 5, the wall of the burner zone was exposed to more CO and less O_2 with a decrease in the burner zone SR. Here, the CO concentration can directly represent the degree of reducing atmosphere. Using a CO mole fraction larger than 0.5% as the criterion of the strongly reducing atmosphere, the difference in each case was within 3% for the OFA ratio up to 30% (C05-S25). Above this value, the area increased rapidly. Therefore, an OFA ratio of 30% can be considered as the limit to avoid excessive corrosion, and 25% would be acceptable, considering the trade-off between NO emission and overall boiler performance.

Table 5. Average mole fractions of CO and O_2 and the area with CO > 0.5% on the wall of the burner zone.

Case		C05-S15	C05-S20	C05-S25	C05-S30	C05-S35
Mole fraction on the burner zone wall [%]	CO	0.329	0.355	0.462	0.609	0.704
	O_2	0.783	0.657	0.617	0.657	0.572
Wall area with CO > 0.5% in the burner zone		16.0	18.7	21.9	35.5	42.1

3.3. *Influence of Different Air Distributions between SOFA and CCOFA*

Figure 10 compares the profiles of gas temperature and concentrations narrowed to the region between burner E level and primary SH for different CCOFA/SOFA distributions with a total OFA ratio fixed at 25% (case set #2 in Table 1). Because the operation conditions of the burner zone were identical, the profiles below E level were the same as those of C05-S20 shown in Figure 7. High CCOFA ratios caused immediate temperature drops from approximately 1510 °C along the four ports, but the fresh air supply recovered the temperature more, by increased oxidation reactions above the CCOFA. It was followed by another temperature drop by the SOFA injection, but the temperature increase above the SOFA was noticeable only at higher SOFA ratios (C10-S15, C05-S20, and C00-S25). The solid carbon concentration above the CCOFA ports was higher for C00-S25 and C05-S20 due to the shortage of O_2, but the rest of the cases had similar values. After the SOFA injection, C00-S25 exhibited more active decrease in solid carbon concentration but its rate was not as fast as those in the burner zone. This led to the highest UBC in the fly ash for this case.

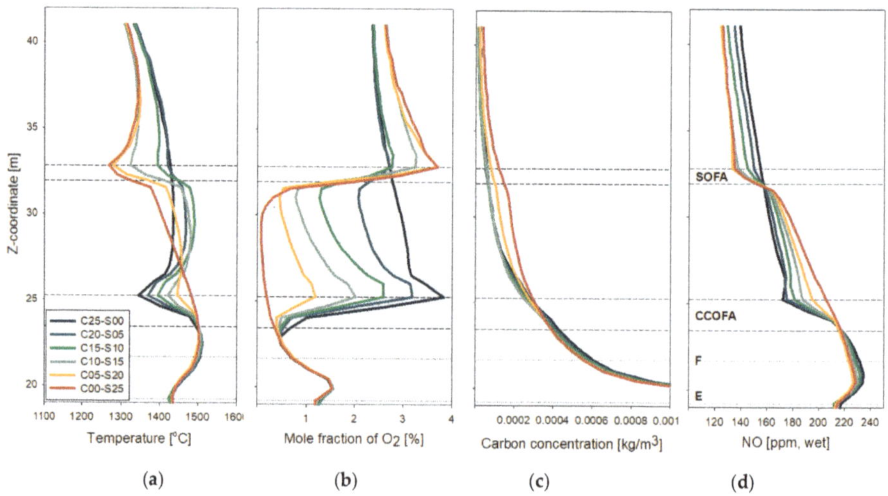

Figure 10. Mass-weighted average profiles of gas temperature and concentrations of O_2, solid carbon, and NO between E burner level and primary superheater for different CCOFA/SOFA distribution (total OFA ratio fixed at 25%). (**a**) Temperature; (**b**) O_2; (**c**) Solid carbon and (**d**) NO.

In the NO profile, the CCOFA injection caused immediate decrease by dilution, splitting the values from 218 ppm on the lowermost CCOFA port level to a range between 174–210 ppm at the top CCOFA port level. The degree of NO reduction reaction in the volume between the CCOFA and SOFA depended on the O_2 concentration, as indicated by the fact that the slope in the NO concentration was proportional to the SOFA ratio. This significantly reduced the difference in the NO concentration before the SOFA level. Then, the trend was almost inverted after the SOFA injection by dilution, but the values of C05-S20 and C00-S25 were similar to that for C10-S15. The NO reduction reactions slowly continued until the gas entered the primary SH and the temperature decreased rapidly.

Figure 11 compares the key performance parameters for different CCOFA/SOFA distributions in case set #2. The UBC content in fly ash, NO emission, and FEGT had favorable results by increasing the proportion of SOFA up to 15% for a total OFA ratio of 25%. Therefore, C10-S15 was the ideal case in this case set. Above this SOFA ratio, both NO emission and FEGT did not change noticeably. However, the UBC content increased rapidly. This was mainly because too strong jets of CCOFA or SOFA led to an inefficient mixing between the char particles and fresh oxygen. This can be confirmed from the path-lines shown in Figure 12 for CCOFA for C25-S00 and SOFA for C10-S15 and C00-S25 drawn on the contours of solid carbon concentration. In the lowermost cross-section, the char particles were present mainly near the wall by the centrifugal force of the fireball. The strong jets by the largest CCOFA or SOFA ratios (C25-S00 and C00-S25, respectively) penetrated deeper into the furnace, and the solid carbon concentration decreased rapidly along their trajectories. However, a significant fraction of char particles escaped the cross-sections through the spaces not covered by the OFA trajectories. In contrast, the jets of CCOFA and SOFA in C10-S15 had a moderate momentum that supplied fresh oxygen to the region near the wall where char particles were more concentrated. This explains the low UBC in this case. The quenching effect by the OFA jets appeared not to be significant, because in all cases, the entrained char was effectively converted. The heat absorption pattern and boiler efficiency were little influenced by the CCOFA/SOFA distribution, as shown in Figure 13.

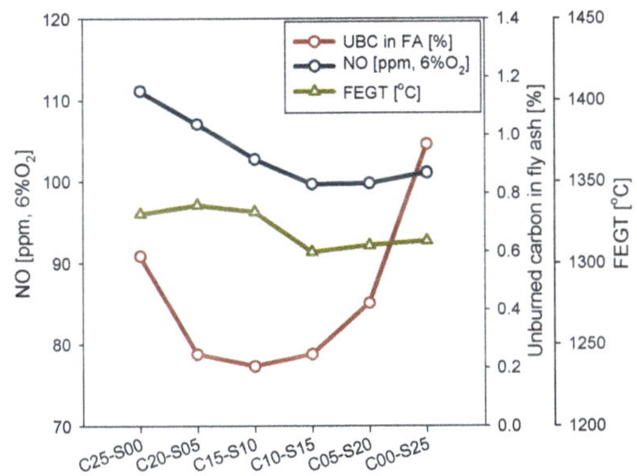

Figure 11. Comparison of NOx emission, UBC in fly ash, and FEGT for different CCOFA/SOFA distributions (total OFA ratio fixed at 25%).

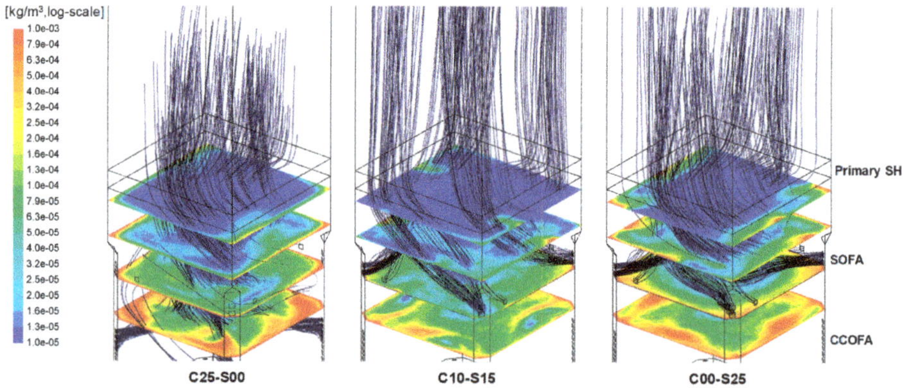

Figure 12. Solid carbon concentration and path-lines of CCOFA (foe C25-S00) and SOFA (for C10-S15 and C00-S25).

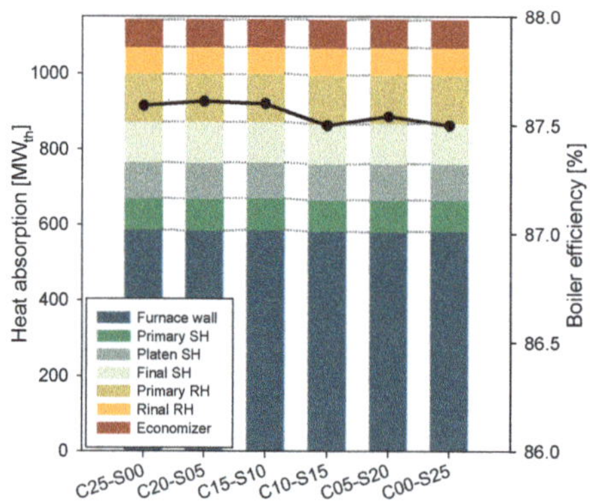

Figure 13. Comparison of heat absorption by heat exchangers and boiler efficiency for different CCOFA/SOFA distributions (total OFA ratio fixed at 25%).

Figure 14 compares the key performance parameters for different CCOFA/SOFA distributions in case set #3 with a total OFA ratio fixed to 30%. The trends in the parameters were similar with those in case set #2 shown in Figure 11, but the NO emissions were lower whereas the FEGT was higher. From the results, C10-S20 could be considered ideal at 30% OFA ratio. Compared to C10-S15 (ideal in case set #2 at 25% OFA ratio), the NO emission (90 ppm) was lowered by 10 ppm whereas the FEGT (1320 °C) and the UBC content (0.39%) were 14 °C and 0.14% higher, respectively.

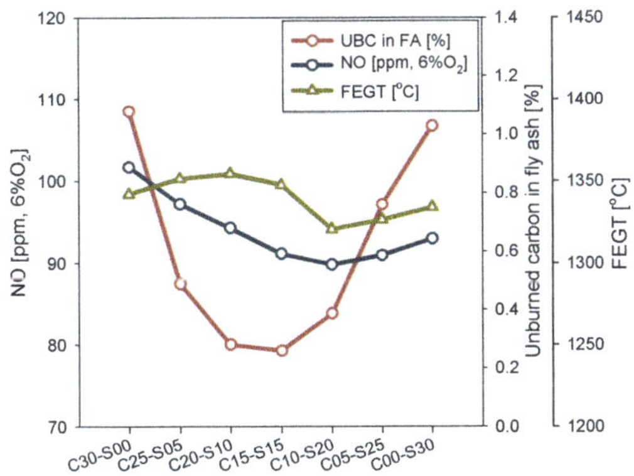

Figure 14. Comparison of NOx emission, UBC in fly ash, and FEGT for different CCOFA/SOFA distributions (total OFA ratio fixed at 30%).

Overall, the results indicate that the CCOFA/SOFA distribution for a fixed total OFA ratio can be optimized for improved boiler performance. Compared to the case with CCOFA only (C25-S00) before the retrofit of this boiler, the overall performance can be significantly enhanced in terms of the NOx emission, the FEGT, and the UBC content in fly ash. With the SOFA installation, cases C10-S15 and C10-S20 were ideal for the OFA ratios of 25% and 30%, respectively. The actual air distribution can be adjusted around these cases, depending on the fuel properties such as ash slagging propensity, fuel N and S contents, heating value, etc.

4. Conclusions

This study investigated the performance of a 500 MWe tangential-firing coal boiler for optimization of the air distribution with an installation of SOFA. Using CFD, the reference case was established in a good agreement with the design data and different air distributions were evaluated for key performance parameters. Increasing the SOFA ratio led to lower NO emission because the NO reduction reactions were more active within the burner zone. However, too large SOFA ratios caused negative impacts on the boiler performance and increased the propensity of slagging and corrosion. With a moderate level of air staging, the NO reduction was also active between the CCOFA and SOFA levels and, therefore, the OFA distribution could be optimized to achieve good boiler performance as well as a low NO emission. For total OFA ratios of 25% and 30% (the burner zone stoichiometric ratio of 0.847 and 0.791, respectively), increasing the SOFA ratio to 15% and 20%, respectively, was ideal for significant reduction in NOx, unburned carbon in fly ash, and furnace exit gas temperature, compared to the case with CCOFA alone before the retrofit. Too high SOFA or CCOFA ratios at the fixed OFA ratios rapidly increased the unburned carbon, because of insufficient mixing between the strong air jets penetrating deep into the center and char particles concentrated near the wall. The heat absorption pattern and boiler efficiency were not noticeably influenced by the distribution between CCOFA and SOFA.

Author Contributions: Mesh generation and tests, H.J. and H.A.; CFD simulations, H.J., K.K., and J.P.; boiler design data and validation of reference case, H.A. and Y.G.; writing—original draft preparation, H.J.; writing—review and editing, C.R.; funding acquisition and project administration, C.R.

Funding: This work was supported by the Korea Institute of Energy Technology Evaluation and Planning (KETEP) affiliated with the Ministry of Trade, Industry and Energy of the Korean Government (Contract No. 20181110200190).

Acknowledgments: The authors would like to thank Doosan Heavy Industries and Construction for their help and technical support.

Conflicts of Interest: The authors declare no conflict of interest.

Nomenclature

Symbols	
A	Pre-exponential factor [s^{-1}]
A_s	Specific tube surface area per unit volume [m^2/m^3]
d	Diameter [cm]
D	Tube diameter [cm]
E	Activation energy [kJ/kmol]
f	Correction factor
f'	Friction factor
G	Mass velocity [$kg/m^2 \cdot s$]
K_{eq}	Chemical equilibrium constant
k	Reaction rate [$g \cdot cm^{-2} \cdot atm^{-1} \cdot s^{-1}$], Thermal conductivity [$W/m \cdot K$]
N	Number of tube
Nu	Nusselt number
P	Pressure [atm]
Pr	Prandtl number
Δp	Pressure difference [atm]
$\dot{q}$	Volumetric heat sink [W/m^3]
R	Universal gas constant, Reaction rate of char [$g \cdot cm^{-2} \cdot s^{-1}$]
Re	Reynolds number
S	Pitch [m]
T	Temperature [K]
V	Volatile matter [kg]
Y	Char core to particle diameter ratio
Greek Symbols	
ε	Porosity of the ash layer, Emissivity
ρ	Density [kg/m^3]
μ	Viscosity [$kg/m \cdot s$]
σ	Stefan-Boltzman constant (5.67×10^{-8} [$W/m^2 \cdot K^4$])
Subscripts	
char	Unreacted char core
conv	Convection
diff	Diffusion rate
dash	Diffusion rate in the ash-layer
gas	Gas
i	Index of gasification reaction
L	Longitudinal
max	Maximum
o	Initial
p	Particle
rad	Radiation
s	Surface
t	Total pressure
T	Transverse

References

1. Kim, H.C.; Kim, S.; Kim, B.-U.; Jin, C.-S.; Hong, S.; Park, R.; Son, S.-W.; Bae, C.; Bae, M.; Song, C.-K.; et al. Recent increase of surface particulate matter concentrations in the Seoul Metropolitan Area, Korea. *Sci. Rep.* **2017**, *7*, 4710. [CrossRef] [PubMed]

2. Choi, J.; Park, R.J.; Lee, H.-M.; Lee, S.; Jo, D.S.; Jeong, J.I.; Henze, D.K.; Woo, J.-H.; Ban, S.-J.; Lee, M.-D.; et al. Impacts of local vs. trans-boundary emissions from different sectors on PM2.5 exposure in South Korea during the KORUS-AQ campaign. *Atmos. Environ.* **2019**, *203*, 196–205. [CrossRef]
3. Nalbandian, H. *NOx Control for Coal-Fired Plant*; CCC/157; IEA Clean Coal Centre: London, UK, 2009.
4. Kang, K.; Ryu, C. Numerical simulation on the effects of air staging for pulverized coal combustion in a tangential-firing boiler. *Korean Chem. Eng. Res.* **2017**, *55*, 548–555.
5. Sankar, G.; Kumar, D.S.; Balasubramanian, K.S. Computational modeling of pulverized coal fired boilers–a review on the current position. *Fuel* **2019**, *236*, 643–665. [CrossRef]
6. Zhou, H.; Mo, G.; Si, D.; Cen, K. Numerical simulation of the NOx emissions in a 1000 MW tangentially fired pulverized-coal boiler: Influence of the multi-group arrangement of the separated over fire air. *Energy Fuels* **2011**, *25*, 2004–2012. [CrossRef]
7. Ma, L.; Fang, Q.Y.; Lv, D.Z.; Zhang, C.; Chen, Y.P.; Chen, G.; Duan, X.N.; Wang, X.H. Reducing NOx emissions for a 600 MWe down-fired pulverized-coal utility boiler by applying a novel combustion system. *Environ. Sci. Technol.* **2015**, *49*, 13040–13049. [CrossRef] [PubMed]
8. Ma, L.; Fang, Q.; Tan, P.; Zhang, C.; Chen, G.; Lv, D.; Duan, X.; Chen, Y. Effect of the separated overfire air location on the combustion optimization and NOx reduction of a 600 MWe FW down-fired utility boiler with a novel combustion system. *Appl. Energy* **2016**, *180*, 104–115. [CrossRef]
9. Zha, Q.; Li, D.; Wang, C.A.; Che, D. Numerical evaluation of heat transfer and NOx emissions under deep-air-staging conditions within a 600 MWe tangentially fired pulverized-coal boiler. *Appl. Therm. Eng.* **2017**, *116*, 170–181. [CrossRef]
10. Qi, X.; Yang, M.; Zhang, Y. Numerical analysis of NOx production under the air staged combustion. *Front. Heat Mass Transf.* **2017**, *8*, 1–8.
11. Zheng, Y.; Gao, X.; Sheng, C. Impact of co-firing lean coal on NOx emission of a large-scale pulverized coal-fired utility boiler during partial load operation. *Korean J. Chem. Eng.* **2017**, *34*, 1273–1280. [CrossRef]
12. Stupar, G.; Tucaković, D.; Živanović, T.; Stevanović, Ž.; Belošević, S. Predicting effects of air staging application on existing coal-fired power steam boiler. *Appl. Therm. Eng.* **2019**, *149*, 665–677. [CrossRef]
13. Yang, W.; Wang, B.; Lei, S.; Wang, K.; Chen, T.; Song, Z.; Ma, C.; Zhou, Y.; Sun, L. Combustion optimization and NOx reduction of a 600 MWe downfired boiler by rearrangement of swirl burner and introduction of separated over-fire air. *J. Clean Prod.* **2019**, *210*, 1120–1130. [CrossRef]
14. Li, S.; Xu, T.; Hui, S.; Wei, X. NOx emission and thermal efficiency of a 300 MWe utility boiler retrofitted by air staging. *Appl. Energy* **2009**, *86*, 1797–1803. [CrossRef]
15. Wu, X.; Fan, W.; Liu, Y.; Bian, B. Numerical simulation research on the unique thermal deviation in a 1000 MW tower type boiler. *Energy* **2019**, *173*, 1006–1020. [CrossRef]
16. Liu, Y.C.; Fan, W.D.; Wu, M.Z. Experimental and numerical studies on the gas velocity deviation in a 600 MWe tangentially fired boiler. *Appl. Therm. Eng.* **2017**, *110*, 553–563. [CrossRef]
17. Park, H.Y.; Baek, S.H.; Kim, H.H.; Kim, Y.J.; Kim, T.H.; Lim, H.S.; Kang, D.S. Reduction of main steam temperature deviation in a tangentially coal-fired, two pass boiler. *Fuel* **2016**, *166*, 509–516. [CrossRef]
18. Xu, L.; Huang, Y.; Zou, L.; Yue, J.; Wang, J.; Liu, C.; Liu, L.; Dong, L. Experimental research of mitigation strategy for high-temperature corrosion of waterwall fireside in a 630 MWe tangentially fired utility boiler based on combustion adjustments. *Fuel Process. Tech.* **2019**, *188*, 1–15. [CrossRef]
19. Jo, H.; Kang, K.; Park, J.; Ryu, C.; Ahn, H.; Go, Y. Detailed assessment of mesh sensitivity and coupled iteration method for CFD simulation of coal combustion in a tangential-firing boiler. *J. Mech. Sci. Technol.* **2019**. under review.
20. ANSYS Inc. *ANSYS Fluent User Guide*; ANSYS Inc.: Cannonsburg, PA, USA, 2016.
21. Niksa, S. *PC Coal Lab Version 4.1: User Guide and Tutorial*; Niksa Energy Associates LLC: Belmont, CA, USA, 1997.
22. Wen, C.Y.; Chaung, T.Z. Entrainment coal Gasification Modeling. *Ind. Eng. Chem. Process Des. Dev.* **1979**, *18*, 684–695. [CrossRef]
23. Jones, W.P.; Lindstedt, R.P. Global reaction schemes for hydrocarbon combustion. *Combust. Flame* **1988**, *73*, 233–249. [CrossRef]
24. Smoot, L.D.; Smith, P.J. *Coal Combustion and Gasification*; Plenum Press: New York, NY, USA, 1985.
25. Magnussen, B.F.; Hjertager, B.H. On mathematical models of turbulent combustion with special emphasis on soot formation and combustion. *Proc. Combust. Inst.* **1977**, *16*, 719–729. [CrossRef]

26. Hanson, R.K.; Salimian, S. Survey of rate constants in H/N/O systems. In *Combustion Chemistry*; Gardiner, W.C., Ed.; Springer: New York, NY, USA, 1984; p. 361.
27. De Soete, G.G. Overall reaction rates of NO and N_2 formation from fuel nitrogen. *Proc. Combust. Inst.* **1975**, *15*, 1093–1102. [CrossRef]
28. Smith, T.F.; Shen, Z.F.; Friedman, J.N. Evaluation of coefficients for the weighted sum of gray gases model. *J. Heat Transf.* **1982**, *104*, 602–608. [CrossRef]
29. Park, H.Y.; Lee, J.E.; Kim, H.H.; Park, S.; Beak, S.H.; Ye, I.; Ryu, C. Thermal resistance by slagging and its relationship with ash properties for six coal blends in a commercial coal-fired boiler. *Fuel* **2019**, *235*, 1377–1386. [CrossRef]
30. Park, J.K.; Park, S.; Kim, M.K.; Ryu, C.; Baek, S.H.; Kim, Y.J.; Kim, H.H.; Park, H.Y. CFD analysis of combustion characteristics for fuel switching to bioliquid in oil-fired power plant. *Fuel* **2015**, *159*, 324–333. [CrossRef]
31. Jakob, M. Heat transfer and flow resistance in cross flow of gases over tube banks. *Trans. ASME* **1938**, *60*, 384–386.
32. Zukauskas, A. Heat transfer from tubes in cross flow. In *Advances in Heat Transfer*; Hartnett, J.P., Irvine, T.F., Jr., Eds.; Academic Press: New York, NY, USA, 1972; Volume 8, pp. 93–160.
33. Kung, S.C. Further understanding of furnace wall corrosion in coal-fired boilers. *Corrosion* **2014**, *70*, 749–763. [CrossRef]

energies

MDPI

Article

Impact of Water Content on Energy Potential and Combustion Characteristics of Methanol and Ethanol Fuels

Jozef Martinka *, Peter Rantuch and Igor Wachter

Faculty of Materials Science and Technology in Trnava, Institute of Integrated Safety, Slovak University of Technology in Bratislava, Jana Bottu 2781/25, 917 24 Trnava, Slovakia

* Correspondence: jozef.martinka@stuba.sk

Received: 12 August 2019; Accepted: 2 September 2019; Published: 10 September 2019

Abstract: Methanol and ethanol are among the most important biofuels and raw materials used to produce biorenewable fuels. These fuels are used with varying water contents. Nevertheless, the exact impact of the water content of these fuels on the energy potential and combustion characteristics is still unknown. Besides that, there are two noticeable risks (environmental impact of combustion and fire risk) associated with their production, processing, and utilization. Likewise, impact of the water content of these fuels on fire risk and the impact of their combustion on the environment is also unknown. The best indicator of energy potential is the effective heat of combustion, and the best combustion characteristic and indicator of the impact of the combustion of alcohols on the environment is the carbon monoxide (CO) yield, whereas the fire risk of liquid fuels is quantified by the flash point and maximum heat release rate (mHRR). The dependency of flash point on the water content was determined via the Pensky-Martens apparatus and the dependencies of the effective heat of combustion, CO yield, and mHRR on the water content were determined via the cone calorimeter. With increased water content, the flash points of both methanol and ethanol exponentially increased and the both effective heat of combustion and mHRR almost linearly decreased. In the range of water content from 0 to 60%, the CO yield of both methanol and ethanol was practically independent of the water content.

Keywords: biorenewable fuels; ethanol; methanol; combustion; effective heat of combustion; carbon monoxide yield; cone calorimeter; alternative fuels; fire risk; flame

1. Introduction

Methanol and ethanol are the primary alcohols utilized as biofuels, biorenewable components of automotive fuels, and raw materials for the production of biorenewable components of automotive fuels. Of the bio-recoverable components of automotive fuels, methyl tert-butyl ether (IUPAC name: 2-methoxy-2-methylpropane, MTBE) is produced from methanol, and ethyl tert-butyl ether (IUPAC name: 2-ethoxy-2-methyl-propane, ETBE) is produced from ethanol. In addition, biomethanol and bioethanol are widely used in households, e.g., as stove fuels.

Both methanol and ethanol are fuels that are fully miscible with water. The unlimited miscibility with water allows for the preparation and utilization of optimal concentration solutions for each application. The best indicator of energy potential of fuels is the effective heat of combustion. The effective heat of combustion of methanol and ethanol fuels decrease with water content increase, but the exact (under real combustion conditions) impact has not been quantified still. Besides that, there are two significant risks associated with the production, processing (e.g., formation of MTBE or ETBE), transport, and utilization of methanol and ethanol. The first major risk is the environmental impact, and the second is the fire risk. Likewise, the exact impact of the water content in methanol and ethanol on the impact of their combustion on the environment and on fire risk has not been quantified.

The impact of alcohol fuel combustion on the environment is expressed by carbon monoxide (CO) yield per unit released heat. The impact of methanol and ethanol combustion on the environment has been partially explored in scientific papers by Cooney et al., Zhang et al., Liang et al., and Melo et al. [1–4]. The fire risk of liquid fuels is determined by their flash point, the heat released, and the CO yield per mass loss. Nevertheless, no scientific paper has examined the complex components of the fire hazard of methanol and ethanol. Selected components related to the fire risk of methanol and ethanol and their mixtures were partially examined by Liaw et al., Vidal et al., and Guo et al. [5–7]. The methanol and ethanol production process is described, e.g., by Kurambhatti et al. and Wang et al. [8,9].

The results from scientific papers of Simonelt et al. and Luche et al. [10,11] show that the combustion of most fuels produce a large quantity of combustion products. To assess the toxicity of combustion products, it is not technically or economically possible to measure and assess all of the components of the combustion products. However, the results of scientific papers of Ometto et al. and Gann et al. [12,13] and IEC 60695-7:2010, IEC 60695-7-3:2011, and C/VM2:2014 technical standards [14–16] show that during the combustion of methanol and ethanol (like with most carbon, hydrogen, and oxygen compounds), only carbon monoxide (CO) is important from the toxicological point of view. Other compounds are formed in too small a quantity and/or their toxicity is too low to significantly affect the resultant toxicity of the combustion products. Additionally, the results from scientific papers of Karlsson and Quintiere [17] and Maricq [18] show that only a negligible amount of soot is produced during the combustion of methanol and ethanol fuels. Therefore, the toxicity of methanol and ethanol combustion products is virtually determined by the amount of CO released.

The impact of water content of methanol and ethanol fuels on their combustion characteristics is very important data for set the parameters of the power equipment (e.g., boilers or stoves) and for fire risk assessment of these fuels. Nevertheless, there is still missing scientific study clarifies impact of water content of methanol and ethanol fuels on their combustion characteristics. Therefore, this study targets to clarify this issue.

The aim of the presented scientific study is to quantify the impact of the water content of methanol and ethanol fuels on the impact of their energy potential, combustion characteristics, the impact of their combustion on the environment, and fire risk.

2. Materials and Methods

Samples of methanol and ethanol fuels with a purity of 99.80% by volume, supplied by Centralchem Slovakia, S.R.O., Bratislava, Slovakia, were used for the research. The raw material for methanol production was wood and the raw material for ethanol production was corn. The samples were examined in both their pure and diluted forms. They were diluted using distilled water to provide water contents of 10, 20, 30, 40, 50, 60, 70, 80, and 90% by volume (pure samples containing 99.80% of the main components are referred to as samples with a water content of 0%).

The samples containing 0, 10, 20, 30, 40, 50, 60, 70, 80, and 90% of water were used to examine the effect of water content on the flash point. To examine the impact of water content on the effective heat of combustion (EHC), heat release rate (HRR), maximum heat release rate (mHRR), maximum average rate of heat emission (MARHE), and CO yield (per unit mass loss and unit released heat), samples containing 0, 20, 40, 60, and 80% of water were used.

The impact of water content on the flash point of methanol and ethanol was determined using the Pensky-Martens apparatus (NPM 131; Normalab, S.A., Naninne, Belgium) according to ISO 2719:2016 [19]. The test procedure complied with ISO 2719:2016 [19] (Test Procedure A) with one modification. The modification of the test procedure consisted of cooling the sample and the test flask of the test apparatus prior to testing. Prior to testing the methanol and ethanol solutions with water contents of 0 to 70%, the samples and the test flask were cooled to −15 °C. Prior to testing the methanol and ethanol solutions with water contents of 80 and 90%, the samples and the test flask were cooled to 0 °C. The Pensky-Martens apparatus has been calibrated according to ISO 2719:2016 [19].

The impact of the water content on the EHC, HRR, mHRR, MRHE, and CO yield (per unit mass loss and unit released heat) was determined using a cone calorimeter (Dual Cone Calorimeter; Fire Testing Technology, Ltd., East Grinstead, UK) according to ISO 5660-1:2015 [20]. The test procedure complied with ISO 5660-1:2015 [20] with one modification. The modification was the replacement of the solid sample holder with a liquid sample container consisting of a steel circular vessel with an internal diameter of 106 mm and a height of 20 mm (the wall thickness of the container was 2 mm). The vessel diameter was chosen so that the surface area of the liquid fuel would correspond to the surface area of the solid fuel required by ISO 5660-1:2015 [20]. The test procedure and the modification were described in detail in the scientific paper of Martinka et al. [21]. During the test, the samples, in compliance with the requirements of ISO 5660-1:2015 [20], were exposed to a heat flux of 10 kw/m^2 from a cone heater (cone heater is a part of Dual Cone Calorimeter; Fire Testing Technology, Ltd., East Grinstead, UK). The sample volume used for each cone calorimeter test was 85 ± 5 mL. The combustion efficiency was calculated from the data measured using the cone calorimeter by using the method according to scientific papers of Ward et al. and Ferek et al. [22,23].

The cone calorimeter has been calibrated in accordance with ISO 5660-1:2015 [20].

The HRR describes amount of heat released in one second from one square meter of the sample. Unit of the HRR is kW/m^2. This unit could also be expressed as kJ/(s × m^2) because kW = kJ/s. The cone calorimeter measure heat release rate for samples with 0.0088 m^2 surface and then recalculate it to 1 m^2 sample surface. The mHRR is defined as the maximum value (peak) of the heat release rate recorded during the test. Another parameter is MARHE. This parameter can be defined as maximum average rate of heat release rate. Units of the mHRR and MARHE are the same as unit of the HRR. The CO yield per mass loss is defined as amount of CO (g) released per 1 kg mass loss of fuel (this parameter express how many CO is released from 1 kg burned fuel). The unit of the CO yield per mass loss unit is therefore g/kg. Similar parameter is the CO yield per released heat. This parameter express amount of CO (g) released per 1 MJ heat released from burned fuel. The unit of the CO yield per released heat is therefore g/MJ.

3. Results and Discussion

The dependence of the flash points of the alcohol–water solutions examined on the water content is shown in Figure 1. Both methanol and ethanol exhibited an exponential flash point dependence on water content. The flash point of the undiluted methanol was 13 °C and the undiluted ethanol had a flash point of 14 °C. In contrast, the flash points of methanol and ethanol with a water content of 90% were 55 °C and 50 °C, respectively. The data shows that the flash points of the alcohols examined exhibited noticeable increases with increasing water content. However, even a water content of 90% cannot be considered as sufficient protection against ignition, and even with such a degree of dilution with water, the examined fuels act as flammable liquids (as evidenced by the fact that their flash points can be determined). Therefore, during their production, processing, and utilization, methanol and ethanol are to be considered as flammable liquids irrespective of their degree of dilution with water.

The impact of the water content on the heat release rate is illustrated in Figure 2. It shows a noticeable decrease in the heat release rate with increasing water content. During combustion, the undiluted alcohols began to boil (under cone calorimeter test conditions). The boiling of undiluted methanol began at approximately 150 s and undiluted ethanol at approximately 200 s (Figure 2). The boiling in Figure 2 shows a rapid increase in the heat release rate to approximately 60 kW/m^2 for methanol or to 150 kW/m^2 for ethanol. For alcohols with a water content of 20%, boiling was only observed for a relatively short period of time (Figure 2). At a water content equal to or above 40%, boiling was not observed. This was due to the lower flame temperature and the resultant decrease in the reverse heat radiation from the flame to the surface of the alcohol–water solution and the higher heat capacity and boiling point of water compared to the alcohols. The reason for the lower flame temperature was that in the combustion zone, apart from alcohol (fuel), water was also evaporated from the alcohol–water solution. As a result, less alcohol was evaporated into the combustion zone (the heat released from the combustion zone reduced), and the heat released from the combustion zone was consumed to heat the combustion products and to heat the evaporated water.

The comparison of the heat release rate of the examined fuels with other scientific papers is rather complicated, as this is the first scientific study to fully examine the impact of water content on heat release rate and the impact of methanol and ethanol combustion on the environment. However, there are scientific papers that have thoroughly examined the heat release rate of other flammable liquid fuels (e.g., [21] examined the fire risk of automotive gasoline). The comparison of Figure 2 with the results of the scientific paper of Martinka et al. [21] shows that methanol and ethanol had a lower maximum heat release rate than automotive gasoline, but the course of the heat release rate was similar. The comparison of the heat release rate for solid biorenewable sources (lignocellulosic fuels) published in scientific papers of Janssens, Maciulaitis et al., Akaki et al., and Carosio et al. [24–27] demonstrates that undiluted methanol and ethanol exhibit higher heat release rates and more even heat release than lignocellulosic fuels. Conversely, methanol with a water content equal to or higher than 40% and ethanol with a water content equal to or higher than 60% showed a lower heat release rate than solid biorenewable fuels according to Janssens, Maciulaitis et al., Akaki et al., and Carosio et al. [24–27].

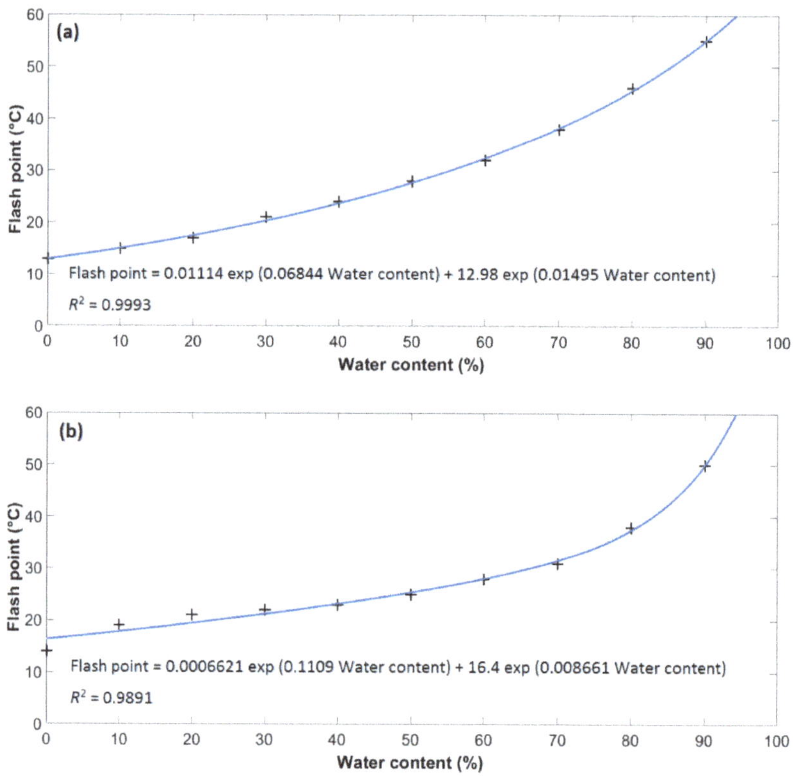

Figure 1. Dependency of the flash point of investigated fuels on the water content: (**a**) Dependency of the flash point of methanol on the water content; (**b**) Dependency of the flash point of ethanol on the water content.

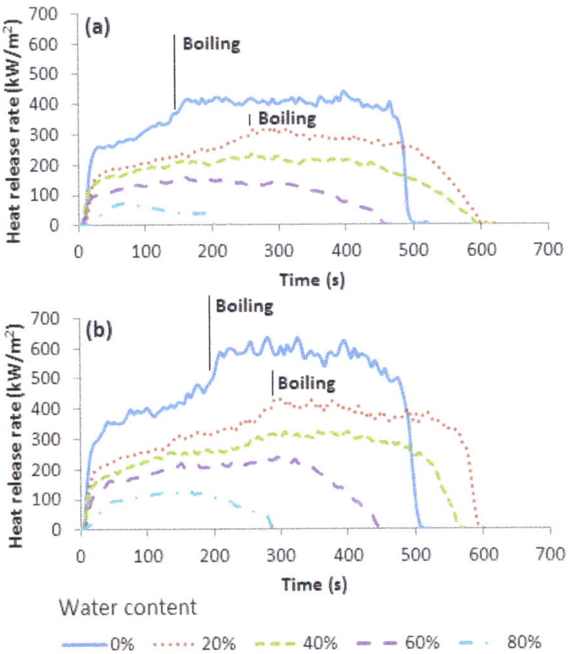

Figure 2. Dependency of the heat release rate of investigated fuels on the water content: (**a**) Dependency of the heat release rate of methanol on the water content; (**b**) Dependency of the heat release rate of ethanol on the water content.

The primary derived heat release rate parameters are the mHRR shown in Figure 3, the MARHE shown in Figure 4, and the EHC shown in Figure 5. The mHRR is the maximum heat release rate (local maximum) recorded during the test. The MARHE was calculated according to Equation (1),

$$\text{MARHE} = \max \frac{\sum_{t_i}^{t} \frac{\text{HRR}(t)+\text{HRR}(t-1)}{2} \cdot {}_{\Delta}t}{t - t_i} \quad (1)$$

where MARHE is maximum average rate of heat emission (kW/m^2), HRR(t) is heat release rate at time t (kW/m^2), HRR(t−1) is heat release rate at time t−1 (kW/m^2), t is time for which average rate of heat emission is calculated (s), t_i is time to ignition (s) and $_{\Delta}t$ is time increment between successive measured values between HRR (s) [28].

The EHC is the amount of heat released per unit mass loss fuel under the conditions of a cone calorimeter test. The mHRR (Figure 3), MARHE (Figure 4), and EHC (Figure 5) linearly decreased with increasing water content.

A comparison of the mHRR (Figure 3) and MARHE (Figure 4) with the data in [24–27,29–32] shows that the undiluted methanol and ethanol both exhibit a higher mHRR and MARHE than most natural polymers (fuels), but are at the same time lower than most synthetic polymers and flammable liquids. This comparison further demonstrates that methanol with a water content of 40% and ethanol with a water content of 60% both exhibit a lower mHRR and MARHE than most substances. For comparison purposes, the combustion and fire characteristics of common alcohols and biorenewable fuels can be found in [33–40].

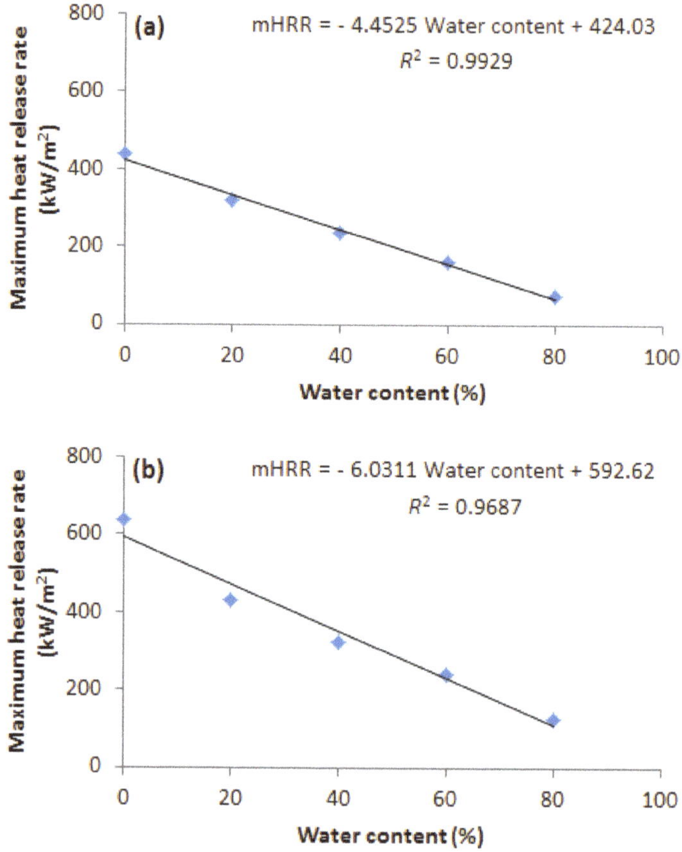

Figure 3. Dependency of the maximum heat release rate of investigated fuels on the water content: (**a**) Dependency of the maximum heat release rate of methanol on the water content; (**b**) Dependency of the maximum heat release rate of ethanol on the water content (mHRR: maximum heat release rate).

The EHC values (Figure 5) of undiluted methanol (23.9 MJ/kg) and ethanol (32.7 MJ/kg) were approximately equal to the maximum theoretical gross calorific values of methanol (22.7 MJ/kg) and ethanol (29.7 MJ/kg). Additionally, they were approximately 20% higher than the maximum theoretical net calorific values of methanol (20.1 MJ/kg) and ethanol (27 MJ/kg) calculated from the standard enthalpy of formation and the vaporization enthalpy of water published in the scientific paper by Haynes [41]. The reason for the slight difference between the gross and net calorific values and the EHC of methanol and ethanol was due to the use of different principles for their measurement. The standard enthalpy of formation needed to calculate the gross and net calorific value is measured in an oxygen calorimeter according to ISO 1716:2018 [42], which measures the heat released based on the increase in water temperature in a calorimetric vessel. The cone calorimeter measures the EHC based on the oxygen consumption (to which 13.1 ± 0.7 kJ of heat per 1 g of consumed oxygen is released during the combustion of most organic substances [43]. In comparison, from the results in [44,45], the net calorific value of most wooden fuels (wood is considered to be a reference material to biofuels comparison) ranges from 17 to 18 MJ/kg. As shown in Figure 5, methanol shows similar EHC values (17.25 MJ/kg) at a water content of 40%, and ethanol also shows similar values (20.4 MJ/kg) at a water content of 60%. This comparison illustrates that from the point of view of EHC, methanol and ethanol show a comparable energy potential as wooden fuels at water contents of 40% for methanol and 60% for ethanol.

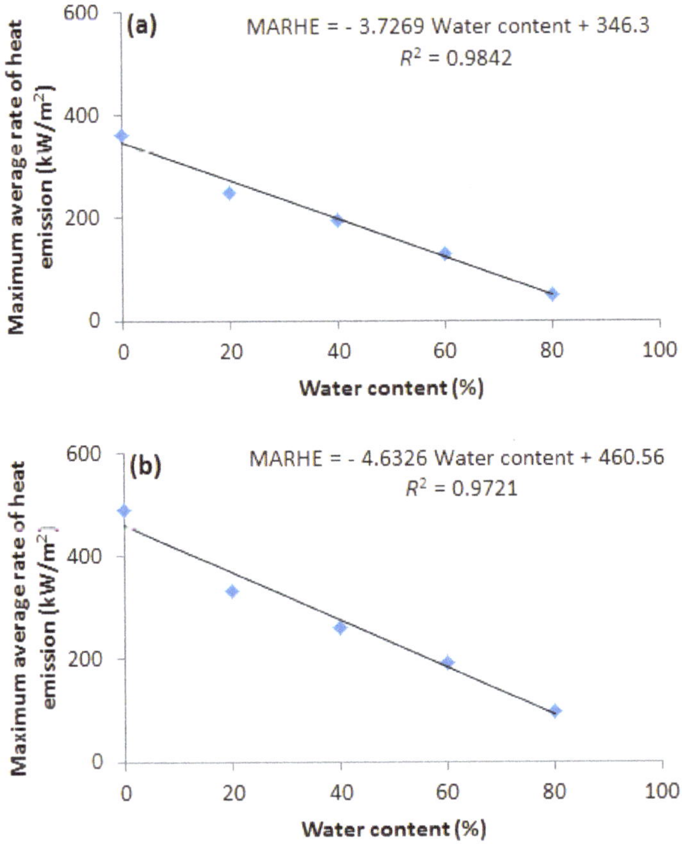

Figure 4. Dependency of the maximum average rate of heat emission of investigated fuels on the water content: (**a**) Dependency of the maximum average rate of heat emission of methanol on the water content; (**b**) Dependency of the maximum average rate of heat emission of ethanol on the water content (MARHE: maximum average rate of heat emission).

Fuel and water were not evaporated from the solution into the combustion zone in the proportions corresponding to their ratio in the solution. This was due to the different boiling temperatures and vaporization enthalpies of alcohols and water. The fuel–water ratio in the combustion zone can be estimated by using the fact that no heat is released from the water evaporated from the solution into the combustion zone. Therefore, the ratio of the EHC for a solution with certain water content to the EHC of pure (undiluted) fuel approximately corresponds to the average fuel–water ratio in the combustion zone. The average fuel–water ratios in the combustion zone for solutions with the water contents examined are shown in Figure 6. Figure 6 shows that the average methanol-to-water ratio, as well as the ethanol-to-water ratio in the combustion zone, decreased almost linearly with increasing water content of the solution. At a water content of 80%, the average methanol-to-water ratio in the combustion zone was approximately 0.50, and the average ethanol-to-water ratio was approximately 0.51 (Figure 6).

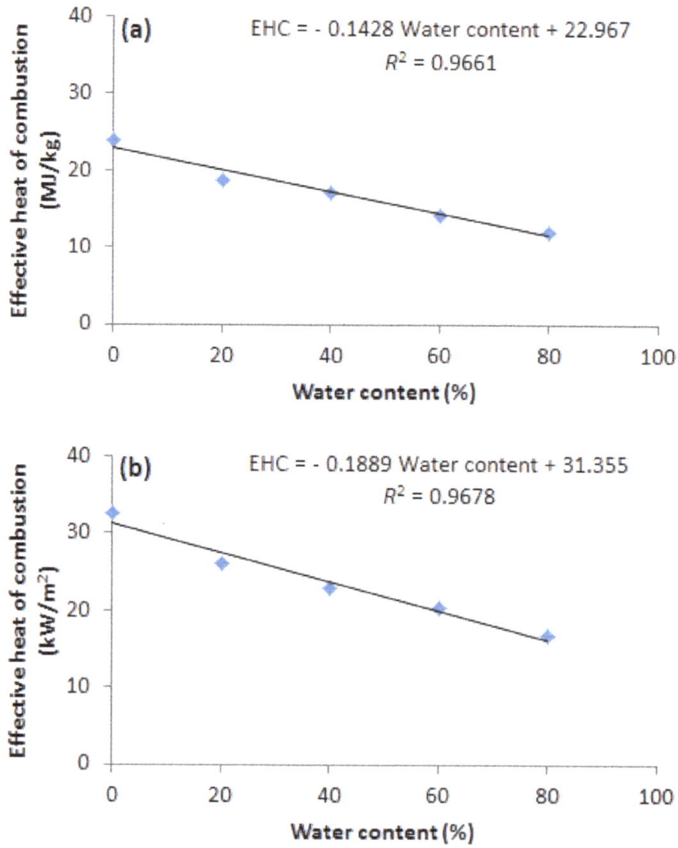

Figure 5. Dependency of the effective heat of combustion of investigated fuels on the water content: (**a**) Dependency of the effective heat of combustion of methanol on the water content; (**b**) Dependency of the effective heat of combustion of ethanol on the water content (EHC: effective heat of combustion).

In addition to the fuel–water ratio (evaporated from the solution) in the combustion zone, another important combustion characteristic is the minimum fuel content in the solution, which is still combustible under the given conditions (critical alcohol content for spontaneous flame extinction). The conditions are represented by the ambient temperature or heat flux exposed to the surface of the combustible liquid. Determination of the critical fuel ratio of the solution (for spontaneous flame extinction) is based on the fact that the EHC of alcohol linearly decreases as water content increases (Figure 5). Thus, if the original mass or volume of the solution and the mass or volume loss of the solution during the spontaneous flame extinction (Table 1) are known, the mass of fuel evaporated from solution can be calculated according to Equation (2) and the mass of water evaporated from the solution can be calculated according to Equation (3),

$$FE = MLS \times (EHC/EPF) \tag{2}$$

$$MEW = (MLS/100) \times (1 - EHC/EPF) \tag{3}$$

where FE is the mass of fuel evaporated (g), MLS is the mass loss of the solution of fuel and water (g), EHC is the effective heat of combustion of a solution of fuel and water (kJ/kg), EPF is the effective heat of combustion of pure fuel (kJ/kg), and MEW is the mass of evaporated water (g).

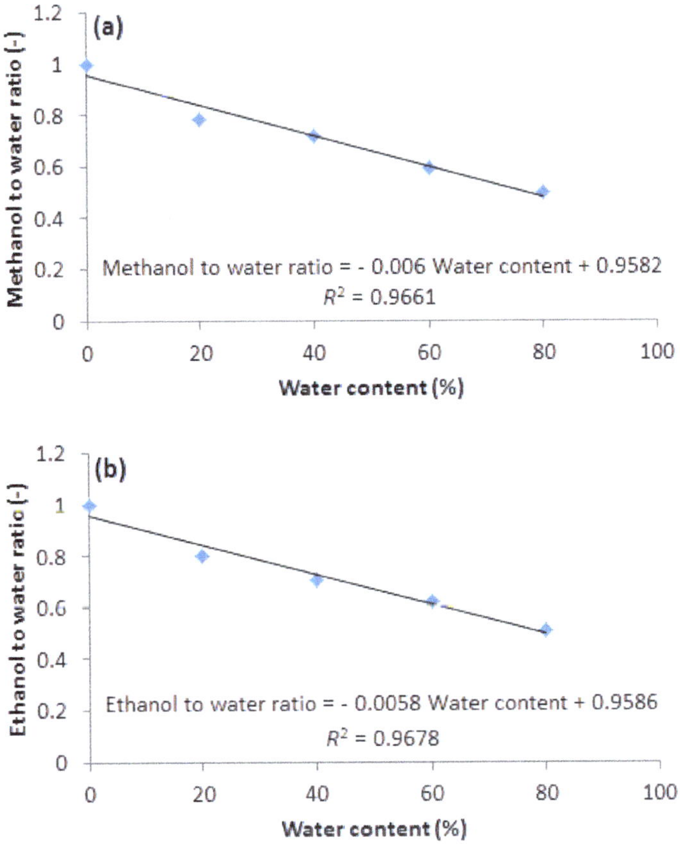

Figure 6. Dependency of the mass ratio of investigated fuels to water (evaporated from solution) in the combustion zone on the water content: (**a**) Dependency of the mass ratio of methanol to water (evaporated from solution) in the combustion zone on the water content; (**b**) Dependency of the mass ratio of ethanol to water (evaporated from solution) in the combustion zone on the water content.

Table 1. Mass loss of fuel solutions from ignition to the spontaneous flame extinction.

Water Content in Solution (%)	60	80
Mass loss of methanol solution (g)	35.5	8.1
Mass loss of ethanol solution (g)	35.4	14.8
Mass of burned methanol (g)	20.9	4.1
Mass of burned ethanol (g)	21.9	7.5
Mass of water evaporated from methanol solution (g)	14.6	4
Mass of water evaporated from ethanol solution (g)	13.5	7.3

The mass of fuel in the solution during spontaneous flame extinction is then calculated as the difference between the fuel in solution before the start of the test and the mass loss of fuel. The critical fuel concentration in the solution is calculated from the mass or volume of fuel in the solution and the volume or mass of the solution (in the case of spontaneous flame extinction). The critical concentrations (mass ratios of fuel to water) for the fuels examined in a water solution, for which spontaneous flame extinction occurs, are illustrated in Table 2. The critical fuel content for a solution with an alcohol content of more than 40% was not determined due the fact that, during the combustion of the 60% and

80% alcohol solutions, only a negligible residue remained (water evaporated into the combustion zone together with the alcohol). The data in Table 2 shows that the critical methanol content (in solution with water) for spontaneous flame extinction is about 15.6 to 17.3%, and the critical ethanol content is about 11 to 14.2%. This conclusion, however, only applies for these experimental conditions. For practical applications, it is necessary to refer to Figure 1, according to which it is possible to determine the flash point of methanol and ethanol even at a water content of 90%. The reason for the difference between the critical water content for spontaneous flame extinction determined using a cone calorimeter according to ISO 5660-1:2015 [20] and the flash point determined using the method according to ISO 2719:2016 [19]) is the different procedures used for the determination of these parameters.

Table 2. Critical fuels contents (ratio of alcohol mass to water mass) in solution for the spontaneous flame extinction.

Water Content in Solution (%)	60	80
Critical methanol content in solution (%)	17.3	15.6
Critical ethanol content in solution (%)	14.2	11

The dependency of the CO yields on the water content in methanol and ethanol is illustrated by Figure 7 (the CO yields per mass loss unit) and Figure 8 (the CO yields per released heat unit). Figure 7 shows that the water content in methanol (in the interval from 0 to 60%) have virtually no effect on the CO yield (per mass loss unit), while the CO yield (per mass loss unit) of ethanol with increasing water content (in the interval of 0 to 60%) decreases. The decrease in the CO yield from ethanol can be explained by the fact that, besides ethanol, water also contributes to the mass loss from the sample solution (input data for the CO yield calculation), and no CO is released in the combustion zone from the water evaporated from the solution. However, this effect did not appear in the decrease of the CO yield (per mass loss unit) from methanol. This is probably due to the lower boiling point of methanol 64.7 °C, according to [41], compared to ethanol 78.4 °C, according to [41], and the faster evaporation of methanol from the water solution. As a result of this faster evaporation, methanol burns in the combustion zone at a higher mass ratio of alcohol to water than ethanol (this conclusion is also confirmed by Figure 6). The result is the independence of the CO yield (per unit of sample mass loss) from methanol to the water content (in the interval 0 to 60%). For both methanol and ethanol with a water content of 80%, a sharp rise in the CO yield (per mass loss) was observed (Figure 7).

The CO yield (per released heat) increases exponentially with increasing water content for both methanol and ethanol (Figure 8). This trend clearly demonstrates the decreased efficiency of combustion of the examined fuels and hence an increasing impact of combustion on the environment with increasing water content. Both fuels showed a maximum CO yield (per both mass loss and released heat) at a water content of 80% (Figures 7 and 8). At the stated water content of the solution, the ratio of fuel to water (evaporated from solution) in the combustion zone is relatively low (Figure 6). A high water level in the combustion zone is manifested through a decrease in flame temperature, through the disruption of the proportion ratio (fuel—oxidant in the combustion zone), which ultimately leads to a reduction in combustion efficiency and a high increase in the CO yields (Figures 7 and 8).

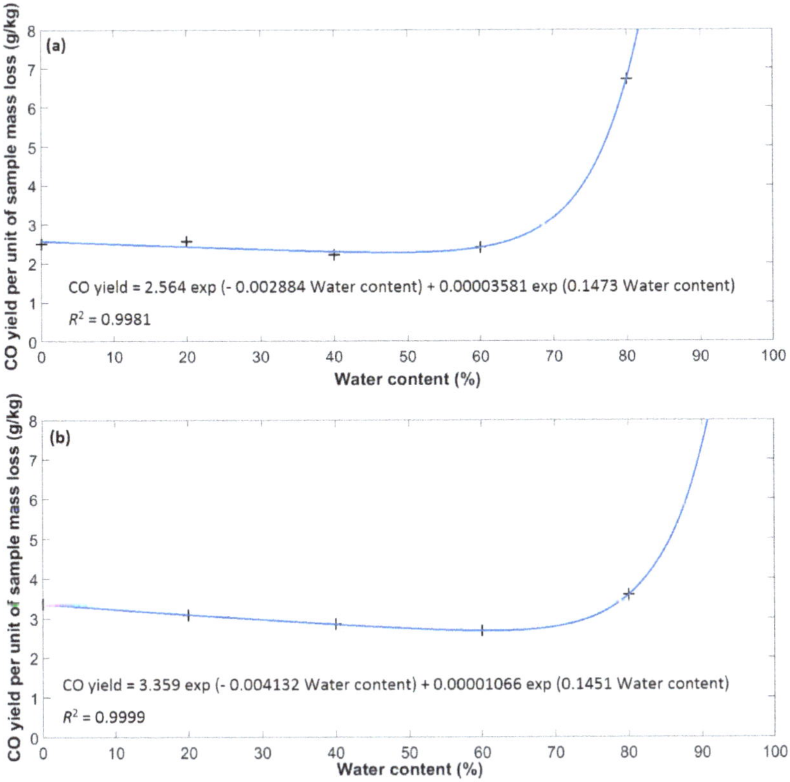

Figure 7. Dependency of the carbon monoxide (CO) yields (per mass loss) of investigated fuels on the water content: (**a**) Dependency of the CO yields (per mass loss) of methanol on the water content; (**b**) Dependency of the CO yields (per mass loss) of ethanol on the water content.

The comparison of the CO yields (per mass loss and released heat) of the examined fuels (Figures 7 and 8) with the CO yields from natural and synthetic polymers (fuels) published in scientific papers by Karlsson and Quintiere [17], DiNenno, and Ozgen et al. [46,47] shows that both methanol and ethanol exhibit a substantially lower CO yield than most natural and synthetic polymers and fuels. Moreover, scientific study by Gardiner [48] demonstrates that the intermediates of burning organic compounds consisting of carbon, hydrogen, and oxygen are oxidized to CO substantially more readily than CO to carbon dioxide (CO_2). Additionally, methanol and ethanol generally only produce a negligible amount of soot. CO is therefore a reliable indicator of the combustion efficiency of methanol and ethanol and the impact of their combustion on the environment. At water contents of 0 to 60%, the impact of the combustion of methanol and ethanol on the environment is negligible due to low CO yields (Figures 7 and 8). At a water content of 80%, the impact of the combustion of methanol and ethanol on the environment, given the higher CO yields (Figures 7 and 8), is more pronounced, but still relatively low in comparison to other substances. Likewise, the contribution of methanol and ethanol fires to the toxicity of combustion products is low due to the low CO yields (Figures 7 and 8) in comparison to other substances.

In addition to CO yields, combustion efficiency is also a reliable combustion characteristic (with the decreasing effectiveness of combustion, the amount of toxic combustion products increases, and more of the substance must burn to obtain the same amount of heat). The impact of water on the combustion efficiency of methanol and ethanol is shown in Figure 9. The combustion efficiency data (Figure 9) roughly corresponds to the inverse data in Figures 7 and 8 and thus confirms the maximum impact of combustion on the environment at a water content of methanol and ethanol of 80%.

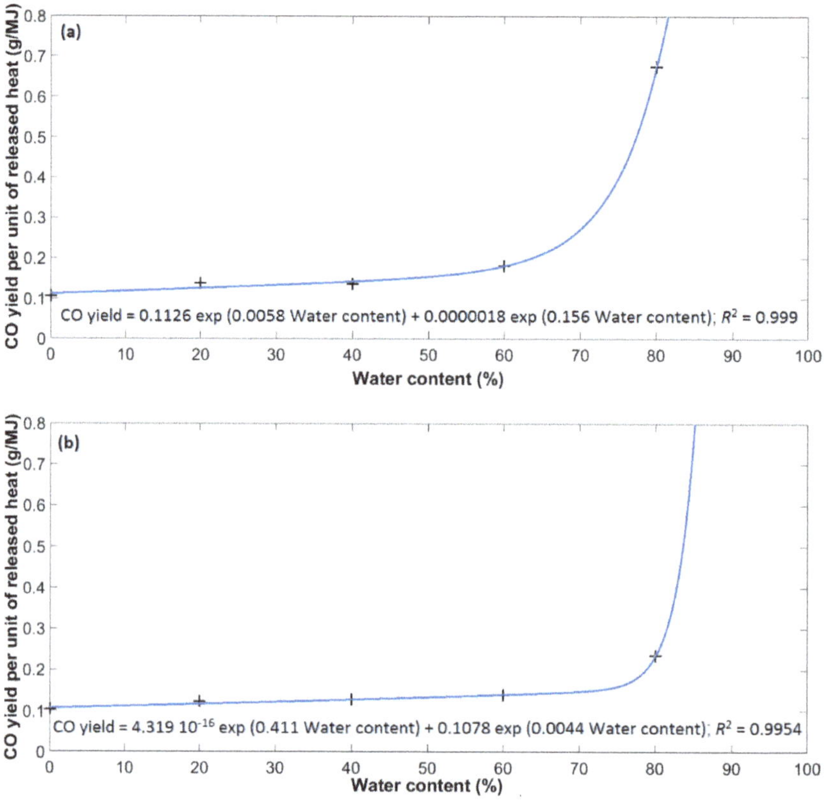

Figure 8. Dependency of the CO yields (per released heat) of investigated fuels on the water content: (**a**) Dependency of the CO yields (per released heat) of methanol on the water content; (**b**) Dependency of the CO yields (per released heat) of ethanol on the water content.

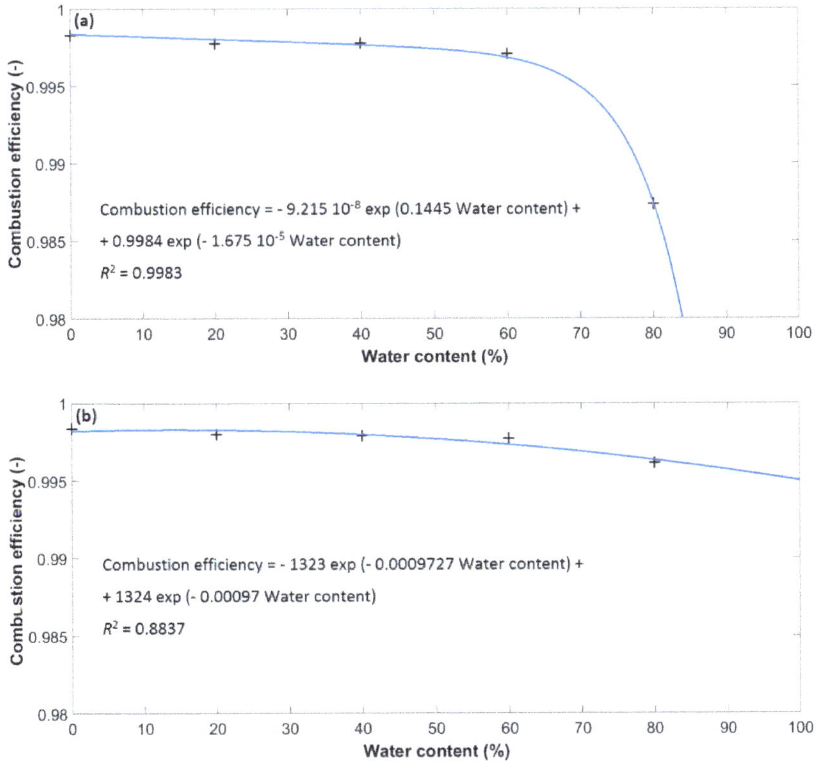

Figure 9. Dependency of the combustion efficiency of investigated fuels on the water content: (**a**) Dependency of the combustion efficiency of methanol on the water content; (**b**) Dependency of the combustion efficiency of ethanol on the water content.

4. Conclusions

This study attempts to quantify the impact of the water content of methanol and ethanol fuels on their energy potential, combustion characteristics, and fire risk. The results of the study may be summarized in the following conclusions.

1. With increasing water content, the flash point of both methanol and ethanol considerably increased. Both methanol and ethanol show an exponential dependence of flash point on water content. Despite the noticeable increase in the flash point (in an undiluted state, methanol and ethanol have flash points of 13 and 14 °C, while for a water content of 90%, the flash point rises to 55 and 50 °C, respectively), the examined alcohols behave as flammable liquids even at a water content of 90%.
2. The EHC, mHRR, and MARHE of methanol and ethanol decrease linearly with increasing water content.
3. The water content of methanol (in the range 0 to 60%) has no practical effect on the CO yield (per mass loss), while the CO yield (per mass loss) of ethanol with decreasing water content (in the range 0 to 60%) slightly decreases. The CO yield (per released heat) of both methanol and ethanol increases with increasing water content. The maximum CO yield (per both mass loss and released heat) of both methanol and ethanol was found to be at a water content of 80%. As the combustion of methanol and ethanol only produces a negligible amount soot, and from a toxicological point of view, CO is the most important combustion product, the impact of the combustion of methanol

and ethanol on the environment is determined only the CO yield (mainly per released heat). Minimizing the impact of methanol and ethanol combustion on the environment can thus be achieved by minimizing the water content.

4. The ratio of methanol to water (evaporated from the solution into the combustion zone) and the ratio of ethanol to water (evaporated from the solution into the combustion zone) both decrease linearly with increasing water content of the solution.
5. A methanol solution with water exposed to a heat flux of 10 kw/m^2 stops burning when the methanol content decreases to 15.6–17.3%. A solution of ethanol with water exposed to the same heat flux stops burning when the ethanol content decreases to approximately 11–14.2 %.

Based on these conclusions, it is possible to state that increase of the water content of the examined fuels has a significant effect on the effective heat of combustion decrease and a moderate effect on combustion efficiency decrease. Based on these conclusions, it is also possible to state that the water content of the examined alcohols has an adverse effect on the environmental impact and fire risk. While most components of fire risk (flash point and released heat) decrease with increasing water content, the environmental impact of combustion (CO yield to released heat) increases as the water content increases.

Author Contributions: Conceptualization, J.M.; Methodology, J.M., I.W., and P.R.; Investigation, P.R. and I.W.; Formal analysis, J.M.; Resources, J.M.; Visualization, J.M.; Writing—original draft, J.M.

Funding: This work was supported by the Slovak Research and Development Agency under the contract No. APVV-16-0223.

Conflicts of Interest: The authors declare no conflict of interest.

Abbreviations

EPF	Effective heat of combustion of pure fuel (kJ/kg)
EHC	Effective heat of combustion (MJ/kg)
ETBE	Ethyl tert-butyl ether (-)
FE	Mass of fuel evaporated (g)
HRR	Heat release rate (kW/m^2)
IUPAC	International Union of Pure and Applied Chemistry (-)
MARHE	Maximum average rate of heat emission (kW/m^2)
MEW	Mass of evaporated water (g)
mHRR	Maximum heat release rate (kW/m^2)
MLS	Mass loss of the solution of fuel and water (g)
MTBE	Methyl tert-butyl ether (-)

References

1. Cooney, C.P.; Worm, J.J.; Naber, J.D. Combustion characterization in an internal combustion engine with ethanol—gasoline blended fuels varying compression ratios and ignition timing. *Energy Fuel* **2009**, *23*, 2319–2324. [CrossRef]
2. Zhang, Z.H.; Cheung, C.S.; Chan, T.L.; Yao, C.D. Experimental investigation on regulated and unregulated emissions of a diesel/methanol compound combustion engine with and without diesel oxidation catalyst. *Sci. Total Environ.* **2010**, *408*, 865–872. [CrossRef] [PubMed]
3. Liang, C.; Ji, C.; Liu, X. Combustion and emissions performance of a DME-enriched spark-ignited methanol engine at idle condition. *Appl. Energy* **2011**, *88*, 3704–3711. [CrossRef]
4. Melo, T.C.C.D.; MacHado, G.B.; Belchior, C.R.P.; Colaco, M.J.; Barros, J.E.M.; De Oliveira, E.J.; De Oliveira, D.G. Hydrous ethanol-gasoline blends-combustion and emission investigations on a flex-fuel engine. *Fuel* **2012**, *97*, 796–804. [CrossRef]
5. Liaw, H.J.; Tang, C.J.; Lia, S.J. A model for predicting the flash point of ternary flammable solutions of liquid. *Combust Flame* **2004**, *138*, 308–319. [CrossRef]

6. Vidal, M.; Rogers, W.J.; Mannan, M.S. Prediction of minimum flash point behaviour for binary mixtures. *Process Saf. Environ.* **2006**, *84*, 1–9. [CrossRef]
7. Guo, Y.; Wei, H.; Yang, F.; Li, D.; Fang, W.; Lin, R. Study on volatility and flash point of the pseudo-binary mixtures of sunflowerseed-based biodiesel + ethanol. *J. Hazard. Mater.* **2009**, *167*, 625–629. [CrossRef]
8. Kurambhatti, C.V.; Kumar, D.; Rausch, K.D.; Tumbleson, M.E.; Singh, V. Ethanol production from corn fiber separated after liquefaction in the dry grind process. *Energies* **2018**, *11*, 2921. [CrossRef]
9. Wang, X.; Martin, N.M.; Nilsson, J.; Carlson, S.; Gustafson, J.; Skoglundh, M.; Carlsson, P.A. Copper-modified zeolites and silica for conversion of methane to methanol. *Catalysts* **2018**, *8*, 545. [CrossRef]
10. Simoneit, B.R.T.; Rogge, W.F.; Mazurek, M.A.; Standley, L.J., Hildemann, L.M.; Cass, G.R. Lignin pyrolysis products, lignans, and resin acids as specific tracers of plant classes in emissions from biomass combustion. *Environ. Sci. Technol.* **1993**, *27*, 2533–2541. [CrossRef]
11. Luche, J.; Mathis, E.; Rogaume, T.; Richard, F.; Guillaume, E. High-density polyethylene thermal degradation and gaseous compound evolution in a cone calorimeter. *Fire Saf. J.* **2012**, *54*, 24–35. [CrossRef]
12. Ometto, A.R.; Hauschild, M.Z.; Roma, W.N.L. Lifecycle assessment of fuel ethanol from sugarcane in Brazil. *Int. J. Life Cycle Assess.* **2009**, *14*, 236–247. [CrossRef]
13. Gann, R.G.; Babrauskas, V.; Grayson, S.J.; Marsh, N.D. Hazards of combustion products: Toxicity, opacity, corrosivity, and heat release: The experts' views on capability and issues. *Fire Mater.* **2011**, *35*, 115–127. [CrossRef]
14. IEC 60695-7-1:2010. *Fire Hazard Testing. Part 7-1: Toxicity of Fire Effluent—General Guidance*, 3rd ed.; International Electrotechnical Commission: Geneva, Switzerland, 2010; p. 50.
15. IEC 60695-7-3:2011. *Fire Hazard Testing. Part 7-3: Toxicity of Fire Effluent—Use and Interpretation of Test Results*, 1st ed.; International Electrotechnical Commission: Geneva, Switzerland, 2011; p. 68.
16. C/VM2:2014. *Verification Method: Framework for Fire Safety Design*, 1st ed.; The Ministry of Business, Innovation and Employment: Wellington, New Zealand, 2014; p. 74.
17. Karlsson, B.; Quintiere, J. *Enclosure Fire Dynamics*, 1st ed.; CRC Press: Boca Raton, FL, USA, 1999; p. 336.
18. Maricq, M.M. Soot formation in ethanol/gasoline fuel blend diffusion flames. *Combust Flame* **2012**, *159*, 170–180. [CrossRef]
19. ISO 2719:2016. *Determination of Flash Point: Pensky-Martens Closed Cup Method*, 4th ed.; International Organization for Standardization: Geneva, Switzerland, 2016; p. 22.
20. ISO 5660-1:2015. *Reaction to Fire Tests. Heat Release, Smoke Production and Mass Loss Rate. Part 1: Heat Release Rate (Cone Calorimeter Method) and Smoke Production Rate (Dynamic Measurement)*, 3rd ed.; International Organization for Standardization: Geneva, Switzerland, 2015; p. 55.
21. Martinka, J.; Chrebet, T.; Balog, K. An assessment of petrol fire risk by oxygen consumption calorimetry. *J. Therm. Anal. Calorim.* **2014**, *117*, 325–332. [CrossRef]
22. Ward, D.E.; Hao, W.M. Air Toxic Emissions from Burning of Biomass Globally: Preliminary Estimates. In Proceedings of the 85th Annual Meeting and Exhibition, Air and Waste Management Association, Kansas City, MO, USA, 16–25 June 1992; Air and Waste Management Association: Kansas City, MO, USA, 1992; pp. 92–98.
23. Ferek, R.J.; Reid, J.S.; Hobbs, P.V.; Blake, D.R.; Liousse, C. Emission factors of hydrocarbons, halocarbons, trace gases and particles from biomass burning in Brazil. *J. Geophys. Res. Atmos.* **1998**, *103*, 32107–32118. [CrossRef]
24. Janssens, M. Rate of heat release of wood products. *Fire Saf. J.* **1991**, *17*, 217–237. [CrossRef]
25. Maciulaitis, R.; Praniauskas, V. Fire tests on wood products subjected to different heat fluxes. *J. Civ. Eng. Manag.* **2010**, *16*, 484–490. [CrossRef]
26. Akaki, T.; Maehara, H.; Tooyama, M. Development of wood and wood ash based hydroxyapatite composites and their fire-retarding properties. *J. Wood Sci.* **2012**, *58*, 532–537. [CrossRef]
27. Carosio, F.; Cuttica, F.; Medina, L.; Berglund, L.A. Clay nanopaper as multifunctional brick and mortar fire protection coating-Wood case study. *Mater. Des.* **2016**, *93*, 357–363. [CrossRef]
28. Zhang, J. Study of Polyamide 6-Based Nanocomposites. Ph.D. Thesis, Polytechnic University, New York, NY, USA, 27 November 2008.
29. Beyer, G. Flame retardant properties of EVA-nanocomposites and improvements by combination of nanofillers with aluminium trihydrate. *Fire Mater.* **2001**, *25*, 193–197. [CrossRef]

30. Wang, S.; Hu, Y.; Li, Z.; Wang, Z.; Zhuang, Y.; Chen, Z.; Fan, W. Flammability and phase-transition studies of nylon 6/montmorillonite nanocomposites. *Colloid Polym. Sci.* **2003**, *281*, 951–956. [CrossRef]
31. Chigwada, G.; Jash, P.; Jiang, D.D.; Wilkie, C.A. Fire retardancy of vinyl ester nanocomposites: Synergy with phosphorus-based fire retardants. *Polym. Degrad. Stab.* **2005**, *89*, 85–100. [CrossRef]
32. Varley, R.J.; Groth, A.M.; Leong, K.H. The role of nanodispersion on the fire performance of organoclay-polyamide nanocomposites. *Compos. Sci. Technol.* **2008**, *68*, 2882–2891. [CrossRef]
33. Srinivasan, K.K.; Mago, P.J.; Krishnan, S.R. Analysis of exhaust waste heat recovery from a dual fuel low temperature combustion engine using an organic Rankine cycle. *Energy* **2010**, *35*, 2387–2399. [CrossRef]
34. Maurya, R.K.; Agarwal, A.K. Experimental study of combustion and emission characteristics of ethanol fuelled port injected homogeneous charge compression ignition (HCCI) combustion engine. *Appl. Energy* **2011**, *88*, 1169–1180. [CrossRef]
35. Liang, C.; Ji, C.; Gao, B.; Liu, X.; Zhu, Y. Investigation on the performance of a spark-ignited ethanol engine with DME enrichment. *Energy Convers. Manag.* **2012**, *58*, 19–25. [CrossRef]
36. Zachar, M.; Mitterova, I.; Xu, Q.; Majlingova, A.; Cong, J.; Galla, S. Determination of fire and burning properties of spruce wood. *Drvna Ind.* **2012**, *63*, 217–223. [CrossRef]
37. Kim, J.; Lee, J.H.; Kim, S. Estimating the fire behavior of wood flooring using a cone calorimeter. *J. Therm. Anal. Calorim.* **2012**, *110*, 677–683. [CrossRef]
38. Pfeil, M.A.; Groven, L.J.; Lucht, R.P.; Son, S.F. Effects of ammonia borane on the combustion of an ethanol droplet at atmospheric pressure. *Combust Flame* **2013**, *160*, 2194–2203. [CrossRef]
39. Cekovska, H.; Gaff, M.; Osvaldova, L.M.; Kacik, F.; Kaplan, L.; Kubs, J. Tectona grandis Linn. and its fire characteristics affected by the thermal modification of wood. *BioRes* **2017**, *12*, 2805–2817. [CrossRef]
40. Martinka, J.; Martinka, F.; Rantuch, P.; Hrusovsky, I.; Blinova, L.; Balog, K. Calorific value and fire risk of selected fast-growing wood species. *J. Therm. Anal. Calorim.* **2018**, *131*, 899–906. [CrossRef]
41. Haynes, W.M. *Handbook of Chemistry and Physics*, 96th ed.; CRC Press: Boca Raton, FL, USA, 2015; p. 2677.
42. ISO 1716:2018. *Reaction to Fire Tests for Products: Determination of the Gross Heat of Combustion (Calorific Value)*, 4th ed.; International Organization for Standardization: Geneva, Switzerland, 2018; p. 30.
43. Thornton, W.M. The relation of oxygen to the heat of combustion of organic compounds. *Philos. Mag.* **1917**, *33*, 196–203. [CrossRef]
44. Gunther, B.; Gebauer, K.; Barkowski, R.; Rosenthal, M.; Bues, C.T. Calorific value of selected wood species and wood products. *Eur. J. Wood Prod.* **2012**, *70*, 755–757. [CrossRef]
45. Todaro, L.; Rita, A.; Cetera, P.; D'Auria, M. Thermal treatment modifies the calorific value and ash content in some wood species. *Fuel* **2015**, *140*, 1–3. [CrossRef]
46. DiNenno, P.J. *SFPE Handbook of Fire Protection Engineering*, 3rd ed.; National Fire Protection Association: Quincy, MD, USA, 2002; p. 1606.
47. Ozgen, S.; Caserini, S.; Galante, S.; Giugliano, M.; Angelino, E.; Marongiu, A.; Hugony, F.; Migliavacca, G.; Morreale, C. Emission factors from small scale appliances burning wood and pellets. *Atmos. Environ.* **2014**, *94*, 144–153. [CrossRef]
48. Gardiner, W.C. *Gas-Phase Combustion Chemistry*, 2nd ed.; Springer: New York, NY, USA, 2000; p. 543.

energies

MDPI

Article

Identification and Quantification of Uncertainty Components in Gaseous and Particle Emission Measurements of a Moped

Barouch Giechaskiel *, Alessandro A. Zardini, Tero Lähde, Michael Clairotte, Fabrizio Forloni and Yannis Drossinos

European Commission – Joint Research Centre, 21027 Ispra, Italy; alessandro.zardini@ec.europa.eu (A.A.Z.); tero.lahde@ec.europa.eu (T.L.); michael.clairotte@ec.europa.eu (M.C.); fabrizio.forloni@ec.europa.eu (F.F.); yannis.drossinos@ec.europa.eu (Y.D.)

* Correspondence: barouch.giechaskiel@ec.europa.eu; Tel.: +39-0332-78-5312

Received: 15 October 2019; Accepted: 13 November 2019; Published: 14 November 2019

Abstract: The recent Euro 4 and 5 environmental steps for L-category vehicles (e.g., mopeds, motorcycles) were mainly designed to reduce the emissions of particulate matter and ozone precursors, such as nitrogen oxides and hydrocarbons. However, the corresponding engine, combustion, and aftertreatment improvements will not necessarily reduce the solid particle number (SPN) emissions, suggesting that a SPN regulation may be necessary in the future. At the same time, there are concerns whether the current SPN regulations of passenger cars can be transferred to L-category vehicles. In this study we quantified the errors and uncertainties in emission measurements, focusing on SPN. We summarized the sources of uncertainty related to emission measurements and experimentally quantified the contribution of each uncertainty component to the final results. For this reason, gas analyzers and SPN instruments with lower cut-off sizes of 4 nm, 10 nm, and 23 nm were sampling both from the tailpipe, and from the dilution tunnel having the transfer tube in closed or open configuration (i.e., open at the tailpipe side). The results showed that extracting from the tailpipe 23–28% of the mean total exhaust flow (bleed off) resulted in a 24–31% (for CO_2) and 19–73% (for SPN) underestimation of the emissions measured at the dilution tunnel. Erroneous determination of the exhaust flow rate, especially at cold start, resulted in 2% (for CO_2) and 69–149% (for SPN) underestimation of the tailpipe emissions. Additionally, for SPN, particle losses in the transfer tube with the closed configuration decreased the SPN concentrations around 30%, mainly due to agglomeration at cold start. The main conclusion of this study is that the open configuration (or mixing tee) without any instruments measuring from the tailpipe is associated with better accuracy for mopeds, especially related to SPN measurements. In addition, we demonstrated that for this moped the particle emissions below 23 nm, the lower size currently prescribed in the passenger cars regulation, were as high as those above 23 nm; thus, a lower cut-off size is more appropriate.

Keywords: particle number; L-category; vehicle regulations; CO_2 emission; tailpipe measurements; dilution tunnel; accuracy; measurement uncertainty; motorcycles

1. Introduction

Air pollution continues to have significant impacts on the economy and the health of the European population, particularly in urban areas. Europe's most serious pollutants in terms of adverse health effects are particulate matter (PM), nitrogen dioxide (NO_2), and ground-level ozone (O_3) [1]. PM is damaging to ecosystems and cultural sites, responsible for reduced visibility, and an important global risk factor for human health. Automotive exhaust PM, due to its important contribution to urban PM, has been subject to progressively more stringent regulations [2].

PM mass emissions are determined gravimetrically by collecting diluted exhaust gas on a filter. In addition to the PM method, the non-volatile (solid) particle number (SPN) emissions are measured by a specified method [3] in the European Union (EU). The SPN sampling is conducted via Constant Volume Sampling (CVS) of the whole exhaust in a dilution tunnel where the exhaust is diluted. The laboratory particle number systems consist of three parts: First, the exhaust gases pass through a hot diluter (≥150 °C), followed by an evaporation tube (350 °C), and lastly by a condensation particle counter (CPC) with 50% counting efficiency at a particle diameter of 23 nm. The methodology and specifications follow the recommendations of the Particle Measurement Program (PMP) [4]. The accuracy of the methodology was estimated to be around 15%. Inter-laboratory correlation exercises, which include the uncertainty of the source (i.e., the car or engine) found a reproducibility of approximately 35% [4,5].

Since 2011, SPN emissions are regulated in the EU for compression ignition (diesel) light-duty vehicles; the limit value is 6×10^{11} particles per km (p/km from now on). Since 2013, on-road compression ignition (diesel) heavy duty-engines are included in the regulations, and since 2014, positive ignition heavy duty-engines with limits of 6×10^{11} p/kWh. A limit of 6×10^{11} p/km was applied to gasoline direct injection vehicles in 2014 [6]. However, for up to three years a particle number emission limit of 6×10^{12} p/km could be applied upon request of the manufacturer. Some non-road engines (power range 19–560 kW), inland waterway vessels (>300 kW), and rail traction engines are regulated for SPN since 2017 with a limit of 1×10^{12} p/kWh. Since 2017, emission testing of light-duty vehicles includes on-road real driving emissions (RDE). Recently, the European Commission decided to extend the lowest detection size of the SPN methodology to 10 nm, to cover cases that have a high fraction of particles in the 10 nm to 23 nm size range (e.g., some vehicles with spark ignition engines) [2].

The only vehicle class so far not covered by SPN limits is the L-category (two- or three-wheel vehicles and quadri-cycles, such as quads and minicars). Nevertheless, moped and motorcycles account about 10% of the passengers mobility fleet [7,8] and contribute significantly to air pollution [9]. Recent EU regulations aim to reduce the share of total road-transport emissions from L-category vehicles as compared to other road vehicle categories with a focus on PM and ozone precursors, such as nitrogen oxides and hydrocarbons. Among other provisions a new test cycle was introduced, the gaseous compounds emission limits were tightened, and a PM mass limit was introduced. In addition, feasibility and cost-effectiveness studies on durability, on-board diagnostics (OBD), in-service conformity, off-cycle emissions, and particle number emissions were required (Regulation (EU) 168/2013). A first investigation on the feasibility and necessity of introducing a SPN limit for L-category vehicles was conducted in 2015 [10]. That research showed that L-category vehicles can have high SPN emissions with a high percentage of them not counted by the current lower limit of 23 nm. However, it was argued that decreasing the lower detectable size could result in volatile artefacts (namely, the formation of volatile particles in the measurement system via nucleation of semi-volatile species—these re-nucleated particles form downstream of the evaporation tube and may be measured by the particle-measuring instrument) and more research was therefore considered necessary. An ad-hoc environmental study concluded that introducing specific SPN limits for L-category vehicles would first require better understanding of the emissions performance of such vehicles, as new emission control technologies at Euro 5 step would become available (from 2020 on) [11]. A detailed analysis of the data concluded that L-category vehicles are a significant contributor to vehicular particulate emissions, and their relative contribution will increase if no measures will be taken [12]. Consequently, a specific particle number limit should be considered for L-category vehicles. A recent study with the latest technology mopeds and motorcycles (all fulfilling the Euro 4 emission standards) reported that, although the PM mass emissions were <1.5 mg/km for all vehicles tested, two motorcycles and the moped were close to the SPN limit for passenger cars and four motorcycles exceeded the limit by a factor of up to four [13]. Even though the repeatability was good (about 10% deviation from the mean), steady state tests with the moped showed big differences between the tailpipe and the dilution tunnel sampling points for sub-23 nm particles. Detailed studies showed that the differences originated mainly from the growth of pre-existing particles with sizes even below 4 nm to the measuring size range of the instruments; however, particles grew due to condensation of desorbed material from the walls of the transfer tube [14,15].

The previous studies suggest that the current legislative SPN methodology cannot be directly extended to L-category vehicles. On the one hand, the mean particle size is close to or lower than the lower size of the current methodology (23 nm) [10], on the other hand sub-23 nm artefacts are more probable with this category of vehicles than with light-duty or heavy-duty category vehicles [16]. Some of the existing facilities are used for both motorcycles and light-duty vehicles and consequently might not be optimized for mopeds. The low exhaust flow rate may lead to high residence times in the transfer lines and consequently high particle losses. Extraction of a sample from the tailpipe for analysis (e.g., real time gas analyzers) (sometimes the term bleed off is used) at the moment is not taken into account in any automation system for SPN emissions (neither in the light-duty regulation). Although this effect might not be important for light-duty vehicles, for mopeds it may result in a significant error. The recently introduced RDE regulations require measurements from the tailpipe. The equivalency of the tailpipe and dilution tunnel locations has been investigated for heavy-duty engines [17] and for light duty vehicles [18], but not for L-category vehicles. The few studies that did some preliminary investigations found big differences both in size distributions and number concentrations [19]. However, there are no studies that have quantified these differences. Thus, there is a need to summarize the open issues for L-category vehicles, especially for particle number measurements, and quantify the measurement uncertainties.

In the International vocabulary of metrology (VIM) [20], Type A evaluation of uncertainty is defined as the method of evaluation of uncertainty by the statistical analysis of series of observations. Repetitions of a test and statistical analysis are classified as Type A uncertainty. Type B evaluation of uncertainty is defined as the method of evaluation of uncertainty by means other than the statistical analysis of series of observations, for instance, calibration reports and guides. This approach was recently followed to estimate the uncertainty of PEMS (portable emissions measurement systems) [21]. In this study we followed a Type B analysis where we tried to collect all parameters that can contribute or influence the final result and quantified them based on theoretical or experimental data. We started from the equations that are used to calculate emissions both at the tailpipe and the dilution tunnel and then discussed possible errors and inaccuracies for each parameter. Even though we did two to three repetitions of each test condition, and thus Type A uncertainty was available, we did not cover the calculation of the combined standard uncertainty and expanded uncertainty, as foreseen in the Guide to the Expression of Uncertainty in Measurement (GUM) [22]. The objective was to identify and quantify the sources of uncertainty/errors that repetitions cannot identify (e.g., tests can be repeatable but erroneous). The focus was on SPN emissions. Gas analyzers and SPN instruments with various cut-off sizes were used to cover current and future regulations. A moped in a facility for motorcycles and passenger cars was chosen as the worst case scenario due to its small weight and engine capacity and the necessity for long transfer lines. Finally, recommendations for appropriate setups and procedures were given.

2. Materials and Methods

2.1. Experimental Setup

The experimental measurements were performed at the Vehicle Emissions Laboratory (VELA 1) of the European Commission's Joint Research Centre. The facility is an emission test cell for L-category vehicles (such as mopeds and motorcycles) and light-duty vehicles (such as small passenger cars). Herein, we consider emission-test measurements performed with a Euro 4 moped with a 50 cm^3 4-stroke engine with a three-way catalyst (>1000 km at the odometer). As shown in Figure 1, which presents schematically the experimental setup, the tailpipe of the moped was connected to a 6 m stainless steel transfer tube. The first section of the transfer tube, the first 4 m, was flexible and heated (heater set to 80 °C), while the second section, the last 2 m, was fixed, insulated, and kept in a climatic room at 20–24 °C. The moped and the first 4 m of the transfer line were inside a different climatic chamber kept at 21–25 °C. The whole exhaust gas was diluted in the dilution tunnel with constant volume sampling (CVS) set to 3.5 m^3/min. With this flow rate, the under-pressure at the tailpipe of the moped is small, and similarly its influence on emissions [23]. The dilution tunnel and the CVS were in the climatic room kept at 20–24 °C.

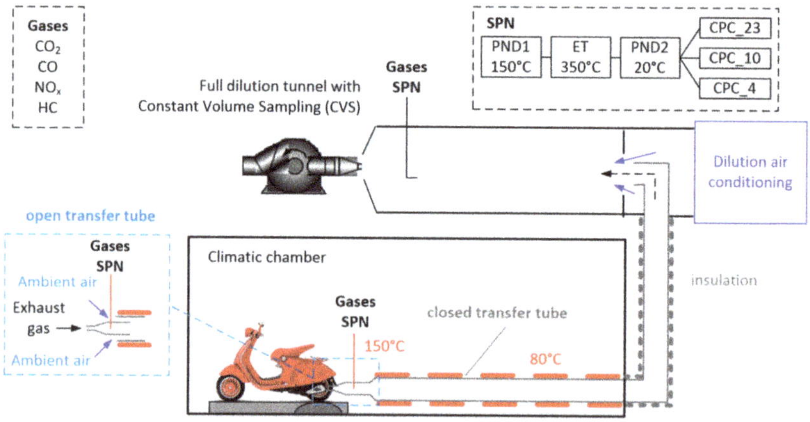

Figure 1. Experimental setup. Gases and SPN were sampled with separate probes at the tailpipe or the dilution tunnel. CPC = Condensation Particle Counter; ET = Evaporation Tube; HC = Hydrocarbons; PND = Particle Number Diluter; SPN = Solid Particle Number.

For gaseous pollutants, two identical AMA i60 (AVL, Graz, Austria) benches were used to measure dry CO and CO_2 with non-dispersive infrared detectors, NO_x with chemiluminescence detectors, and total hydrocarbons (HC) with flame ionization detectors. One system was connected to the tailpipe and the other to the full dilution tunnel with CVS.

The particle measurement systems we used in this study were AVL particle counters (APC 489, Graz, Austria) [24], which consisted of the following parts: A hot diluter at 150 °C, an evaporation tube at 350 °C, a cold diluter with filtered ambient air at 20 °C, and three Condensation Particle Counters (CPCs). As shown in Figure 1, the CPCs had different 50% counting efficiencies: At 23 nm (model 3790 from TSI, USA), at 10 nm (model 3772 from TSI, USA), and at 4 nm (model 3752 from TSI, USA) [25,26]. Two such SPN measurement systems were used. They were identical and freshly calibrated. One (SPN #1) measured always at the tailpipe and the other (SPN #2) at the dilution tunnel (CVS). The SPN #1 was connected to the tailpipe with a stainless steel 0.5 m heated tube at 120 °C.

The transfer tube that conducts the exhaust flow from the moped tailpipe to the dilution tunnel was usually connected to the moped. For some tests, however, the connection was left open on the side normally connected to the moped (Figure 1, left inset). The tailpipe instruments were sampling at the moped tailpipe, before the opening. The CVS under-pressure sucked ambient air also from the opening in addition to that from the external line for main dilution. The dilution factor of this first dilution is not known; based on some steady state tests it is expected to be around 2:1 at the maximum speed and higher at lower speeds [14,19]. This configuration (open transfer tube) is allowed in the EU motorcycles/mopeds regulation, and it has been used previously [23]. We note, however, that the closed configuration is more common [7,27–32].

The test cycle was the Worldwide harmonized Motorcycle Test Cycle (WMTC) class 1 consisting of a cold engine start phase followed by the same phase in hot engine conditions for a total of 7.6 km and 45 km/h maximum speed [33].

2.2. Theoretical Background

L-category regulation provides the equations for gaseous pollutants measured from the dilution tunnel with bags. In our case, we used the integrated signal from the analyzers in order to compare real time signals as well. Equations from the tailpipe can be found in passenger cars regulations at the RDE (Real-Driving Emissions) regulation or at the heavy-duty engines regulation and were summarized elsewhere for gaseous pollutants [34]. Here, the equations for SPN will be given for both the tailpipe and the dilution tunnel, because there is lack of information and it is the most complex case (see also [18]). In addition, we will expand them to take into account the extracted flow from the tailpipe (bleed off). Other topics that can influence the results and have been mentioned in the literature will also be shortly presented.

2.2.1. Solid Particle Number (SPN) Emissions

The tailpipe instantaneous particle number flow rate $SPN_{TP,i}$ [p/s] can be determined by multiplying the instantaneous particle number concentration $C_{TP,i}$ [p/m^3] (normalized to 0 °C and 101.3 kPa) by the instantaneous exhaust mass flow rate $Q_{exh,i}$ [kg/s] (aligned to each other), and by dividing with the density of the exhaust gas ρ_{exh} [kg/m^3] at 0 °C. A density of 1.2931 kg/m^3 was used for gasoline (E10) (Regulation (EU) 2017/1154). The total tailpipe emissions SPN_{TP} [p/km] were calculated by integrating the $SPN_{TP,i}$ rate over the test cycle and dividing by the distance covered D [km] (7.6 km in our tests).

$$SPN_{TP,i} = C_{TP,i}\ Q_{exh,i}/\rho_{exh} \tag{1}$$

$$SPN_{TP} = \Sigma\ SPN_{TP,i}/D \tag{2}$$

When a flow is extracted from the exhaust gas, the equation is still valid if the Q_{exh} includes this extracted flow rate.

The CVS instantaneous particle number flow rate $SPN_{CVS,i}$ [p/s] was calculated as above using the CVS flow rate Q_{CVS} [kg/s] and the density of air ρ_{air} [kg/m^3] at 0 °C. The total tailpipe emissions SPN_{CVS} [p/km] were calculated by integrating the $SPN_{CVS,i}$ rate over the test cycle and dividing by the distance covered D [km].

$$SPN_{CVS,i} = C_{CVS,i}\ Q_{CVS,i}/\rho_{air} \tag{3}$$

$$SPN_{CVS} = \Sigma\ SPN_{CVS,i}/D \tag{4}$$

When a flow is extracted from the dilution tunnel, the equation is still valid if the Q_{CVS} includes this extracted flow rate; however, the correction is usually negligible.

The final SPN_{CVS} result is equivalent to the one in the passenger cars and trucks regulations. The time alignment is less critical in this case:

$$SPN_{CVS} = C_{CVS,ave}\ V_{CVS}/D \tag{5}$$

where V_{CVS} [l] is the volume of the diluted exhaust gas (normalized to 0 °C and 101.3 kPa) and $C_{CVS,ave}$ [p/m^3] is the mean concentration during the cycle duration (normalized to 0 °C and 101.3 kPa) (Regulation (EU) 2017/1151).

The particles extracted from the tailpipe $SPN_{extr,i}$ were calculated using the instantaneous particle number concentration $C_{TP,i}$ [p/m^3] (normalized to 0 °C and 101.3 kPa) and the instantaneous flow rate of the instruments sampling from the tailpipe $Q_{instr,i}$ [kg/s]. In our case, the total extracted flow rate was constant and equal to 19–23 l/min due to the SPN instrument(s) and the gas analyzers. The total particle emissions extracted from the tailpipe SPN_{extr} [p/km] were calculated by integrating the $SPN_{extr,i}$ rate over the test cycle and dividing by the distance covered D [km].

$$SPN_{extr,i} = C_{TP,i}\ Q_{instr,i}/\rho_{exh} \tag{6}$$

$$SPN_{extr} = \Sigma\ SPN_{extr,i}/D \tag{7}$$

The corrected CVS results $SPN_{CVS,corr,i}$ were calculated by adding the SPN emissions of the dilution tunnel $SPN_{CVS,i}$ and the SPN extracted from the tailpipe $SPN_{extr,i}$.

$$SPN_{CVS,corr,i} = SPN_{CVS,i} + SPN_{extr,i} \tag{8}$$

For completeness, another way to calculate the SPN emissions due to the extracted flow from the tailpipe, when no tailpipe SPN measurements are available, is to recalculate every second of the concentrations at the dilution tunnel by using the "true" dilution ratio at the CVS as if no flow had been extracted. The reason is that the extraction of a flow from the tailpipe results in higher dilution in the dilution tunnel, which is not considered in the calculations. This is equivalent to increasing the exhaust flow that would enter the dilution tunnel by the flow extracted by the instruments connected to the tailpipe. As an example, using the difference between CVS total (Q_{CVS}) and dilution air flow (Q_{air}) rates as a proxy of the exhaust flow, the final SPN concentration would be:

$$SPN_{CVS,corr,i} = SPN_{CVS,i}/(Q_{CVS} - Q_{air,i}) \times (Q_{CVS} - Q_{air,i} + Q_{instr,i}) \tag{9}$$

Details about other methods determining the exhaust flow rate or the dilution factor are given in the next paragraph.

For the open transfer tube case, the same tailpipe equations can be used as long as the tailpipe SPN measurements are taken before the dilution at the tailpipe and the exhaust flow rate is estimated correctly.

The results of the SPN emissions at the tailpipe need correct time alignment between the SPN signal and the exhaust flow rate. Misalignment of a few seconds can have a significant effect [35]. Here, the effect of misalignment was checked by moving the SPN signal for ±1 second. Long sampling lines might result in signal deformation due to mixing and diffusion and need advanced mathematical models to "reconstruct" the original signal [36–38]. The tailpipe measurements due to the short sampling lines should not have this issue. It should be mentioned though that the deformation of the signal from the engine to the tailpipe is not of importance in this paper, because only tailpipe with dilution tunnel results are compared [39].

2.2.2. Exhaust Flow Rate

The exhaust flow rate Q_{exh} can be determined with various methods. The direct measurements with exhaust flow meters are sensitive to the exhaust gas fluid conditions, such as the temperature, flow rate, composition, and pressure [40]. In small one- or two-cylinder gasoline engines, there is also a problem with pulsating flows especially during idling, which can even include reverse flow direction. For this reason, today, there are no exhaust flow meters for mopeds and small motorcycles and indirect methods are used. In the case of non-road, small utility purpose engines, for example, the sum of intake air mass and fuel mass flow is used as the exhaust mass flow [41]. Alternatively, one of these two flows can be calculated from the other with the measurement of the air/fuel ratio obtained with a lambda sensor.

Here, we used three methods applicable in the laboratory: The instantaneous difference between total CVS flow Q_{CVS} (corrected for any flows extracted from the dilution tunnel) and dilution air flow (Q_{air}) (Equation (10)), the CVS flow and the instantaneous dilution factor (DF) calculated by the CO_2 tracer method (Equation (11)), or the carbon balance method (CB) (Equation (12)).

$$Q_{exh,air,i} = Q_{CVS} - Q_{air,i}, \tag{10}$$

$$Q_{exh,CO2,i} = Q_{CVS}/DF_{CO2,i}, \quad (11)$$

$$Q_{exh,CB,i} = Q_{CVS}/DF_{CB,i}, \quad (12)$$

$$DF_{CO2,i} = C_{CO2,TP,i}/C_{CO2,CVS,i}, \quad (13)$$

$$DF_{CB,i} = 13.4/(C_{CO2,CVS,i} + C_{CO,CVS,i} + C_{HC,CVS,i}) \quad (14)$$

where C is the concentration [%] of the pollutant in the dilution tunnel (CVS) or tailpipe (TP) for each second i. The concentrations should refer to the wet exhaust (i.e., the dry-to-wet correction should be applied [42]) and should also be corrected for the dilution air background. The dry-to-wet correction is important for the tailpipe concentrations, while the dilution air background correction is important for the dilution tunnel concentrations. The 13.4 is the theoretical CO_2 concentration for a gasoline engine, around 13.4%, which is obtained with the average composition of gasoline fuel (C_8H_{15}) [40]. Note also that the CO_2 tracer method needs a CO_2 analyzer sampling from the tailpipe beyond that at the dilution tunnel, and this extracted flow has to be considered in the calculations. The extracted flow rate has to be added to the equations above to find the actual exhaust flow rate.

$$Q_{exh,i} = Q_{exh,method,i} + Q_{instr,i} \quad (15)$$

2.2.3. Cold Start

During cold start condensation can take place and this has an effect up to 10% on the results of instruments measuring on dry basis (e.g., CO_2 or CO) [42]. According to our knowledge there are no SPN instruments measuring in dry basis. Therefore, the correction is not needed.

2.2.4. Particle Dynamics

The main aerosol processes from the tailpipe to the full dilution tunnel are agglomeration, diffusion, and thermophoresis [18,43,44].

Agglomeration arises from particle to particle collisions where the colliding particles are attached together, but retain their identity and shape. Agglomeration changes the number concentration and size of particles, but not the total particle mass concentration. The extent of agglomeration depends on the initial particle concentration and, to a smaller degree, on particle size and surrounding temperature.

Diffusion is the net movement of particles (or gas molecules) from high to low concentration. Smaller particles diffuse faster than larger particles. Therefore, diffusion is most important for particles smaller than about 50 nm in diameter. Particle concentration decreases in the exhaust transfer lines due to diffusional deposition of particles on the surfaces.

Thermophoresis is the motion of a particle in a temperature gradient. Due to the higher rate of collisions of gas molecules on the hotter side the particle, thermophoresis results in particle movement towards the colder side (i.e., particles in high temperature exhaust gas to the colder walls).

Other transport losses (e.g., gravitational deposition, inertial impaction, electrophoretic) and sampling losses (due to anisoaxial and anisokinetic extraction of sample) were considered negligible as they affect larger particles than those emitted of the moped of this study (<1 μm) [44]. A detailed presentation of the topics is out of the scope of this paper. Details can be found in the literature [18,43,44].

This study focused on measurements of solid particles carried out following the current EU legislation (Regulation (EU) 2017/1151) and on the related uncertainty. The measurements of total particles (including volatiles) are not included in the legislation, both in the laboratory and on the road (Regulation (EU) 2017/1154). They depend on many parameters, such as the dilution ratio and the dilution air temperature, which can result in orders of magnitude difference between different settings [45]. This topic has been discussed in the literature [46] and was beyond the scope of this study.

2.2.5. Artefacts

When comparing solid particle number concentrations, it has to be ensured that no artefacts take place. The volatile artefact is the formation, downstream of the evaporation tube, of volatile particles via re-nucleation of semi-volatile species in the measurement system [16,47]. These particles are volatiles but are counted as solids. The appearance of solid particles in the transfer line due to formation or growth in the transfer tube is defined as a nonvolatile artefact [14,15]. Artefacts at sub-23 nm particle measurements of mopeds and motorcycles have been reported in the literature [13,15]. The topic will not be further discussed; it should be noted though that due to the low exhaust gas temperatures and the high employed dilution in the SPN systems, we do not expect any artefacts in our results.

3. Results

3.1. Exhaust Flow

Figure 2 plots the exhaust flow rate of the motorcycle as estimated by the three methods (Equations (10)–(12)). There is a very good agreement between the different methods. Even though not shown here, the carbon balance method without the CO_2 dilution air correction would slightly overestimate the exhaust flow rate. The CO_2 tracer method was considered the basis for further evaluations.

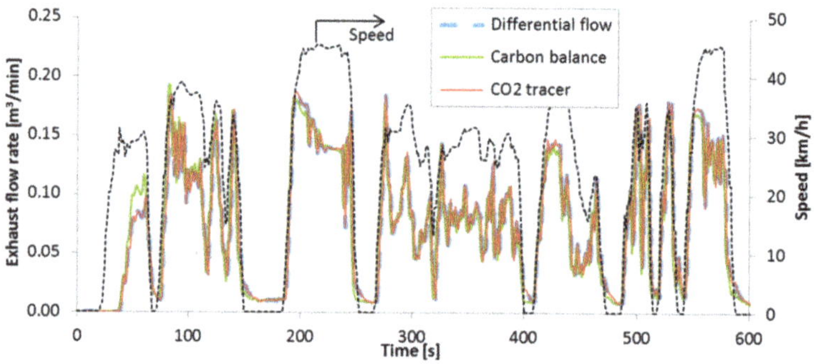

Figure 2. Exhaust flow rate as determined by three methods for the first 600 s of the Worldwide harmonized Motorcycle Test Cycle (WMTC) (closed transfer tube) (Equations (10)–(12)). The extracted flow (0.019 m^3/min) should also be added.

At a next step, the calculated exhaust flow rate with the CO_2 tracer method was plotted for the two configurations with closed and open transfer tube (Figure 3) during the first 250 s of a cold start WMTC. After the first minute, there is a generally good agreement within the experimental uncertainty. However, there is a significant difference at the beginning of the cycle: The exhaust flow rate with the transfer tube closed starts much later. The reason is that there is a big "dead" volume in the transfer tube that delays the CO_2 signal at the dilution tunnel. The gas transport delays, mixing and dispersion phenomena, are responsible for smoothing and distortion of the original emission signal [36,38,39]. Furthermore, even though the transfer tube is heated, there may still be colder spots where condensation can take place. This condensation will be different with the open and closed configurations, resulting in different actual CO_2 concentrations. However, the dry-to-wet corrections of the CO_2 signal do not take this into account [42]. We cannot exclude completely the possibility that the moped behaved differently with open and closed configurations. However, the CO_2 tailpipe signals (and the results later) did not show any difference. This figure clearly shows the importance of having short transfer lines. In the following results, the first 50 s of the calculated exhaust flow rate were corrected using the exhaust flow rate of the open configuration; the effect of this correction will also be shown.

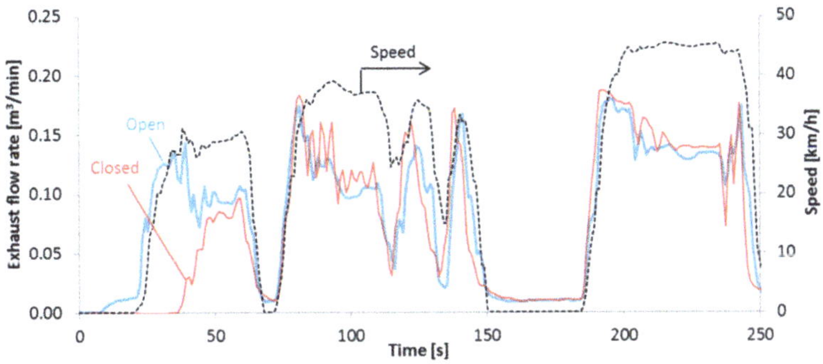

Figure 3. Comparison of the exhaust flow estimated with the CO_2 tracer method during cold start in open/closed configuration. The speed profile is from the open configuration test. The extracted flow (0.019 m^3/min) should also be added.

3.2. CO_2 and Gaseous Pollutants

Figure 4 plots the CO_2 emissions over the WMTC as determined at the tailpipe and the dilution tunnel with integration of the gas analyzers signals with the transfer tube open and closed. The agreement is very good within the experimental uncertainty. For the tailpipe measurements, the improvement of the exhaust flow determination at the first minute of cold start (as discussed in Figure 3) increased the CO_2 emissions by 2.3%. However, single factor analysis of variance (ANOVA) showed that the CO_2 differences were not statistically significant with or without the correction for the first 50 s of the of exhaust flow ($p = 0.23$, $N = 10$). The dilution tunnel results were increased 24% to 31% from the addition of the mass of the extracted exhaust from the tailpipe sampling (Equation (7)). This percentage was similar to the mean extracted flow rate; the extracted 19–23 l/min corresponded to 23–28% of the mean total exhaust flow rate.

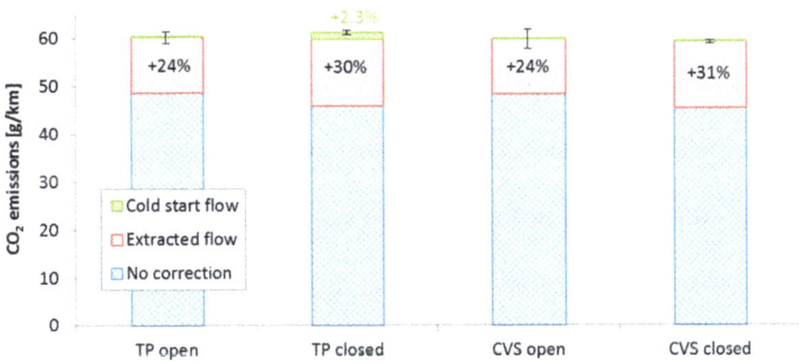

Figure 4. CO_2 emissions at the tailpipe (TP) or the dilution tunnel (Constant Volume Sampling (CVS)) with open or closed transfer tube over an entire WMTC test (blue areas). The corrections due to the extracted tailpipe flow are also given (red areas). The additional emissions by correcting the first 50 s of the exhaust flow rate based on the open configuration (Figure 3) is given for the TP closed case (green dashed area). Error bars show one standard deviation of 2 (open) to 3 (closed) repetitions.

For the rest gases, the results were similar: The extracted flow rate from the tailpipe contributed 19% to 30%, while the first 50 s erroneous exhaust flow rate determination affected 3.5% (CO), 23% (NO_x), and 51% (hydrocarbons) of the rest gases. Thus, the contribution was pollutant specific.

3.3. Solid Particle Number (SPN)

Figure 5 plots the SPN real time emissions over the tested cycle (WMTC) for the two configurations (open and closed transfer tube). With the open configuration (Figure 5, upper panel), the SPN at the CVS and tailpipe (TP) agree very well, with the exception of the first minute, where the CPC of the tailpipe particle number system was saturated (it measured around 60,000 p/cm^3 where the maximum is around 10,000 p/cm^3). The 10 nm was saturated for approximately 25–30 s, while the 23 nm CPC for 10–15 s. The good agreement between tailpipe and dilution tunnel results indicates that there were no artefacts (i.e., formation of particles after the tailpipe). This was expected because the exhaust gas temperature was low (<200 °C) and the instruments dilutions were high (>1000:1).

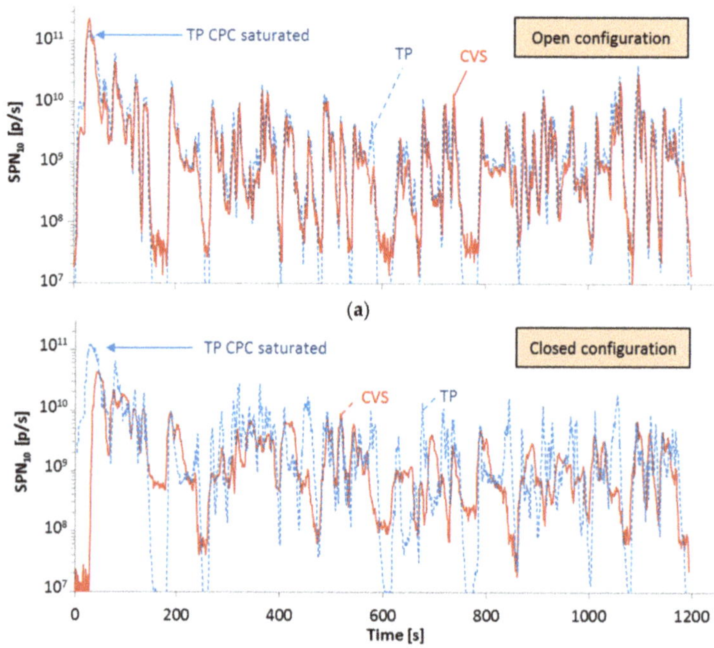

Figure 5. Real time solid particle number emissions (SPN) >10 nm with (**a**) open (upper panel) or (**b**) closed (lower panel) transfer tube. Measurements at the tailpipe (TP) and the dilution tunnel (CVS) are separately plotted for each case.

The picture is very different with the closed configuration (Figure 5, lower panel). The tailpipe results are similar to the open configuration (compare blue dashed TP lines of lower and upper panels), as expected, because the sampling location was the same. However, the SPN at the dilution tunnel is smoother and in some cases the particle peaks appear later due to the variable residence time in the transfer tube. The concentrations seem lower, especially in the first minute. The emissions are very high at cold start, leading to considerable particle losses mainly due to agglomeration. These losses yield decreased particle concentrations at the dilution tunnel.

Figure 6 summarizes the results of all tests conducted: Tailpipe and dilution tunnel measurements with open and closed configurations. The results are given separately for the three CPCs with lower particle sizes of 4 nm, 10 nm, and 23 nm. The CVS results were corrected for the extracted flow rate from the tailpipe (Equation (7)) (bleed off), while for the TP results the extracted flow rate was taken into account by adding it to the exhaust flow rate. The "bleed off" correction was 19–26% for the open transfer tube case and 39–73% for the closed one. The percentages depend on the absolute emission levels, and also on when the emissions take place. The percentages are higher with the closed configuration because the absolute levels are lower. The tailpipe results with the first 50 s corrected exhaust flow rate for the cold start (Figure 3) were 69% to 149% higher than without the correction due to the high SPN emissions at the cold start. After correction, the tailpipe measurements with the closed configuration were close to the tailpipe measurements with the open configuration.

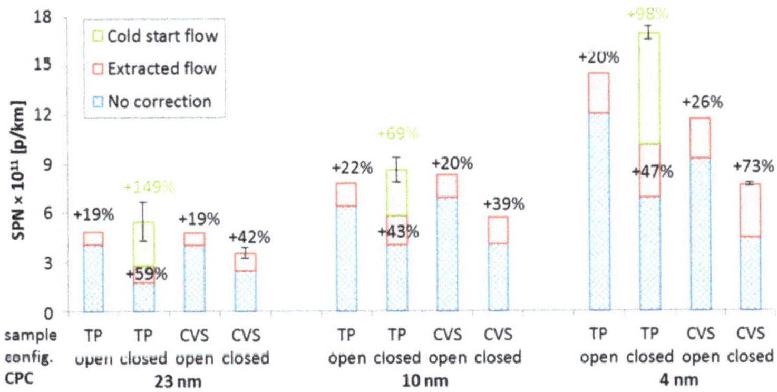

Figure 6. Solid particle number emissions (SPN) at the tailpipe (TP) or the dilution tunnel (CVS) with open or closed transfer tube (blue areas). The corrections due to the extracted tailpipe flow are also given (red area). The additional emissions by correcting the first 50 s of the exhaust flow rate based on the open configuration (Figure 3) is given for the TP closed case (green area). Error bars show one standard deviation of 2–3 repetitions.

After corrections, and assuming that the tailpipe results with the open configuration are the reference values as best available option, the differences become –7% to +4% for the tailpipe with the closed configurations, –2% to +19% for the dilution tunnel with the open configuration, but –19% to –41% for the dilution tunnel with the closed configuration. The lower closed configuration results indicate 20–40% particle losses in the exhaust transfer line. Theoretical calculations using the 4 nm concentrations and a mean size of 20 nm estimated the contribution of agglomeration around 30% [44]. The contribution of thermophoresis was estimated to be <2%, and of diffusion <5%. As the corrected results were in good agreement to each other, the possibility of sub-23 nm artefacts is unlikely in our tests.

Based on the open transfer tube results, the specific moped had SPN_{23} emission levels below the light-duty vehicles limit set for particles >23 nm (6 × 10^{11} p/km), but higher when considering the particles below 23 nm. Based on the increasing emissions with decreasing lower size, the solid particle size distribution peaked at a size close to 10 nm.

4. Discussion

This study, based on transient tests of a moped, quantified the uncertainties in emissions measurements using gaseous analyzers and solid particle number (SPN) instruments with cut-off sizes of 4 nm, 10 nm, and 23 nm at the tailpipe and the dilution tunnel. Two configurations were tested: Closed and open transfer tube to the dilution tunnel; both allowed in the current EU regulations (for gaseous pollutants). The results of this study can be used to draft a list of issues to be discussed in current and future legislation development and research activities. This is the first study to discuss uncertainties of this type for L-category vehicles. We are also not aware of similar discussions for passenger car and trucks. Although the results cannot be directly generalized to other mopeds and motorcycles (relative magnitude of the uncertainties may differ), Table 1 identifies common issues which will be of significance.

Table 1. Contribution of various parameters on results.

	TP Open	TP Closed	CVS Open	CVS Closed	Comments
Exhaust flow	Y	Y	N	N	Depends on the method. In this study ±5%
Extracted flow	Y	Y	Y	Y	The effect can be assumed similar to the extracted flow (CO_2) or higher (SPN) (>25% in this study)
Dead volume of transfer line	N	(Y)	N	Y	Will affect exhaust flow determination, (depending on the method) due to gas mixing and dispersion. Effect on SPN >50% in this study
Concentrations	Y	Y	N	N	Cold start concentrations might exceed max range of tailpipe instruments
Particle losses	N	N	N	Y	Not for gases. Agglomeration for concentrations >1 × 10^7 p/cm^3; 30% effect for peaks >1 × 10^8 p/cm^3
Misalignment	Y	Y	N	N	For ±1 s, in this study around ±5%

The results had to be corrected for the extracted flow rate from instruments sampling from the tailpipe. This is typically done by the laboratory automation software for gaseous pollutants, but not for SPN. The typical approach is to have two instruments measuring simultaneously at tailpipe and dilution tunnel and add to dilution tunnel results the particles that are extracted from the tailpipe. Using the CVS real time signal and correcting with the "true" CVS dilution factor based on the differential method (Equation (9)) would be another approach when no tailpipe measurements are taken. This correction was on the order of 19–73% in this study due to the low exhaust flow rate of the moped (extracted flow 23–28% of mean flow). Thus, the instrument used (SPN instrument and gas analyzers) were taking a significant part of the moped's exhaust flow. The Global Technical Regulation (GTR) 15 for light-duty vehicles allows a maximum 0.5% of the exhaust flow to be extracted and not returned in order to validate a type-approval test. Practically, for this moped it would require instruments with flow rates of <0.25 L/min, which do not exist. Thus, the extracted flow correction might be necessary if in the future measurements from the tailpipe (e.g., for additional pollutants or PEMS) will be required.

The tailpipe measurements need the exhaust flow rate for the calculation of the emissions, which can introduce an error. We used three methods, all applicable only in the laboratory (not on-road): Difference of CVS total and dilution air flow rates, CVS total flow rate, and dilution factors based on CO_2 tracer or carbon balance. The agreement of the three methods was excellent (5%). For diesel vehicles, big errors have been reported with the CO_2 tracer method during fuel cut-offs [48], but for stoichiometric gasoline engines, the error should be small because the CO_2 concentration remains constant over most operating conditions. More studies are needed to better determine the exhaust flow associated uncertainty, e.g., with other methods, such as intake air and fuel consumption or the lambda measurement. This will be even more important when direct methods (i.e., exhaust flow meters) will appear in the market. The time misalignment of the analyzers' signals and the exhaust flow rate signals for the tailpipe measurements can have an effect. For SPN, the effect of ±1 s misalignment was ±5%, reaching ±10% in a few cases, which was in agreement with another study for light-duty vehicles [18]. Another study found up to 20% effect on the estimated catalyst efficiency for a ±2 s misalignment [48]. The time misalignment of the CO_2 could also introduce an error on the exhaust flow determination with the CO_2 tracer method. An important issue was that with closed configuration the exhaust flow of the first 50 s was erroneously calculated to be zero due to the "dead" volume of the transfer tube and the associated dispersion and diffusion phenomena. This led to an underestimation of around 69% to 149% in the SPN results, but less than 2.3% for CO_2. The error was significant for SPN because the majority of the SPN emissions take place during cold start. Thus, this error will be important for engines from which the majority of the emissions take place at the beginning of the test (cold start). It could be avoided with better determination of the exhaust flow (e.g., exhaust flow meter or measurement of intake air and fuel flow). For SPN measurements, the instrument dilution (2000:1) that was used was not enough for the 23 nm and 10 nm CPCs during cold start (10–30 s). This resulted in an uncertainty in the tailpipe SPN results. Comparing with the concentrations of the CPCs at the CVS with the open configuration for those seconds, the underestimation of the emissions was negligible for the 23 nm CPC but around 10% for the 10 nm CPC. Higher dilutions can solve this issue, but higher dilutions have higher uncertainty [24]. Thus, the tailpipe method needs further investigation in the future before introducing it in the regulation.

For the specific moped and tests, the applied corrections increased the emissions significantly. For example, for >10 nm, the tailpipe and dilution tunnel with closed configuration emissions were around 4×10^{11} p/km; with the corrections, the emissions were approximately 112% (tailpipe) and 39% (dilution tunnel) higher. The tailpipe results with the two configurations (open and closed transfer tube) and the dilution tunnel results with the open configuration were within ±10% for the 23 nm and 10 nm CPCs and ±20% for the 4 nm CPCs. The results of the closed configuration were 30% (23 nm, 10 nm) to 47% (4 nm) lower due to particle losses in the transfer tube (30% from agglomeration, <5% from diffusion, and <2% from thermophoresis). Decrease of particle levels of 10–40% have been reported for passenger cars [18,43]. The open configuration decreased the agglomeration and thermophoretic losses due to the immediate cooling dilution at the tailpipe and due to the decreased residence time in the exhaust transfer line. There were no indications that the open or closed configurations formed particles, but this needs more detailed studies as the partitioning of the semi-volatile particles will be different for the two configurations [30]. Tests with heavy-duty diesel and CNG (compressed natural gas) engines with cold and hot dilution showed similar results with 23 nm and 10 nm CPCs. Thus, we expect small effect if any also for mopeds and motorcycles [49]. Dedicated studies have shown that particles may be formed or grow to the measurement range of instruments from the silicone, Teflon parts, or desorbed material of the transfer tube when the exhaust gas temperature is high. These phenomena typically lead to quite large errors that in some cases reach an order of magnitude difference [14,16]. The open configuration seems to minimize such artefacts, because the exhaust gas temperature drops to low levels (<150 °C) [14]. It should be mentioned though that the uncertainty of the SPN instruments are in the 10–15% range [3], and thus the errors associated with the extracted flow, the cold start, and the particle losses are significant. The lower results from the dilution tunnel with the closed configuration shows that L-category studies so far might have been underestimating the emissions [10,13]. The exact magnitude depends on the exhaust gas temperatures, the size of particles, and the SPN concentration levels. The major contributor seems to be the agglomeration losses during cold start where the concentration levels are high. One of the key messages of this comparison is that for future SPN regulations the open transfer tube configuration is more appropriate. In order to avoid the uncertainties related to the open configuration (i.e., how much open, the ambient air might not be filtered, etc.), it is the authors' opinion that the open configuration could be replaced with a mixing tee; a commonly used approach for gasoline light-duty vehicles. In this case, the flow of the dilution air entering is well controlled and, in addition, the air is filtered minimizing the contribution of ambient particles to the results. Special attention should be given to the material of the tube after the mixing tee in order to minimize particle losses.

Closing, it should be mentioned that the SPN_{23} emissions after corrections were around 5×10^{11} p/km; lower than the current light-duty vehicle limit of 6×10^{11} p/km. However, the SPN_{10} and SPN_4 emissions were 8.3×10^{11} p/km and 14×10^{11} p/km, respectively, indicating that the current methodology captures only a small percentage of the emitted particles. This is in agreement with other studies for mopeds and motorcycles [10,13] and references therein. It also raises concerns when comparing SPN emissions using instruments with different lower sizes. The most important message though, based also on the previous studies [10,13], is that future regulations for mopeds and motorcycles should consider a size lower than the current of 23 nm. Similar discussion is on-going for light-duty and heavy-duty vehicles, where high sub-23 nm particle fractions (>100%) have been measured in some cases, for instance, motorcycles [10], port-fuel injection engines [6], compressed natural gas engines [18], and heavy-duty engines [50,51].

5. Conclusions

This study summarized the issues that have been raised in the literature regarding solid particle number (SPN) measurements of L-category vehicles (e.g., mopeds and motorcycles), namely: Lower particle size that should be counted, possibility of artefacts, use of long lines from the vehicle to the dilution tunnel in combination with low exhaust flow rates, extraction of a sample from the tailpipe for analysis, exhaust flow rate determination, time alignment, and cold start. In this study we quantified the uncertainties in emission measurements of a moped, with emphasis on SPN and CO_2 and concluded on the setup with the highest accuracy for regulatory purposes.

The dilution tunnel measurements needed a 24–31% correction for CO_2 and 19–73% for SPN to account for exhaust extracted upstream of the dilution tunnel from instruments sampling from the tailpipe (in our case 23–28% of the mean total exhaust flow) (bleed off). Furthermore, the particles losses in the transfer tube from the moped to the dilution tunnel decreased the SPN emissions by an additional 30% (dilution tunnel vs. tailpipe). The exhaust flow was determined via the difference between total and dilution air flow, the CO_2 tracer method, and the carbon balance. In this study, we estimated about 5% of uncertainty associated to the average exhaust flow, which propagates to the final emission results. The wrong estimation of the exhaust flow during the first minute of the cold start, due to the "dead" volume in the transfer tube to the dilution tunnel, resulted in a >50% error in SPN. A time misalignment of ±1 second gave differences of the SPN emissions on the order of ±5%.

Based on the results of this study, dilution tunnel SPN measurements in open configuration without additional instruments sampling from the tailpipe would give the most accurate results for regulatory purposes in terms of particle emissions. Tailpipe measurements for regulatory purposes need further investigations when specific exhaust flow meters will be available.

The specific moped had SPN emissions of 5×10^{11} p/km: Lower than, but close to the passenger cars limit when considering particles larger than 23 nm. The emissions were double when smaller particles were included and exceeded the limit indicating that for this category a lower cut-off size is more appropriate. With high dilution ratios (around 1000:1) and the tailpipe in open configuration, and consequently, lower exhaust temperatures, the possibility of particle losses due to agglomeration or formation of particles due to high temperatures can be minimized.

Author Contributions: Conceptualization, B.G.; formal analysis, B.G.; writing—original draft preparation, B.G.; writing, review and editing, A.A.Z., T.L., M.C., F.F., and Y.D.

Funding: This research received no external funding.

Acknowledgments: The authors would like to acknowledge D. Lesueur, P. Le Lijour, and M. Sculati for the experimental support.

Conflicts of Interest: The authors declare no conflict of interest.

Disclaimer: The opinions expressed in this manuscript are those of the authors and should in no way be considered to represent an official opinion of the European Commission. Mention of trade names or commercial products does not constitute endorsement or recommendation by the authors or the European Commission.

References

1. Guerreiro, C.; González Ortiz, A.; de Leeuw, F.; Viana, M.; Colette, A.; European Environment Agency. *Air Quality in Europe—2018 Report*; European Environment Agency: Luxembourg, 2018; ISBN 978-929-213-989-6.
2. Giechaskiel, B.; Lähde, T.; Suarez-Bertoa, R.; Clairotte, M.; Grigoratos, T.; Zardini, A.; Perujo, A.; Martini, G. Particle number measurements in the European legislation and future JRC activities. *Combust. Engines* **2018**, *174*, 3–16.
3. Giechaskiel, B.; Mamakos, A.; Andersson, J.; Dilara, P.; Martini, G.; Schindler, W.; Bergmann, A. Measurement of automotive nonvolatile particle number emissions within the European legislative framework: A review. *Aerosol Sci. Technol.* **2012**, *46*, 719–749. [CrossRef]

4. Giechaskiel, B.; Dilara, P.; Andersson, J. Particle measurement programme (PMP) light-duty inter-laboratory exercise: Repeatability and reproducibility of the particle number method. *Aerosol Sci. Technol.* **2008**, *42*, 528–543. [CrossRef]
5. Giechaskiel, B.; Casadei, S.; Mazzini, M.; Sammarco, M.; Montabone, G.; Tonelli, R.; Deana, M.; Costi, G.; Di Tanno, F.; Prati, M.; et al. Inter-laboratory correlation exercise with portable emissions measurement systems (PEMS) on chassis dynamometers. *Appl. Sci.* **2018**, *8*, 2275. [CrossRef]
6. Giechaskiel, B.; Joshi, A.; Ntziachristos, L.; Dilara, P. European regulatory framework and particulate matter emissions of gasoline light-duty vehicles: A review. *Catalysts* **2019**, *9*, 586. [CrossRef]
7. Iodice, P.; Senatore, A. Exhaust emissions of new high-performance motorcycles in hot and cold conditions. *Int. J. Environ. Sci. Technol.* **2015**, *12*, 3133–3144. [CrossRef]
8. European Commission, Directorate General for Mobility and Transport. *EU Transport in Figures: Statistical Pocketbook*; European Commission: Luxembourg, 2019; ISBN 978-92-76-03843-6.
9. Platt, S.M.; Haddad, I.E.; Pieber, S.M.; Huang, R.J.; Zardini, A.A.; Clairotte, M.; Suarez-Bertoa, R.; Barmet, P.; Pfaffenberger, L.; Wolf, R.; et al. Two-stroke scooters are a dominant source of air pollution in many cities. *Nat. Commun.* **2014**, *5*, 3749. [CrossRef]
10. Giechaskiel, B.; Zardini, A.; Martini, G. Particle emission measurements from L-category vehicles. *SAE Int. J. Engines* **2015**, *8*, 2322–2337. [CrossRef]
11. Ntziachristos, L.; Vonk, W.A.; Papadopoulos, G.; van Mensch, P.; Geivanidis, S.; Mellios, G.; Papadimitriou, G.; Steven, H.; Elstgeest, M.; Ligterink, N.E.; et al. *Effect Study of the Environmental Step Euro 5 for L-Category Vehicles*; European Commission: Luxembourg, 2017; ISBN 978-92-79-70203-7.
12. Kontses, A.; Ntziachristos, L.; Zardini, A.A.; Papadopoulos, G.; Giechaskiel, B. Particulate emissions from L-Category vehicles towards Euro 5. *Env. Res.* **2019**.
13. Giechaskiel, B.; Zardini, A.A.; Lähde, T.; Perujo, A.; Kontses, A.; Ntziachristos, L. Particulate emissions of Euro 4 motorcycles and sampling considerations. *Atmosphere* **2019**, *10*, 421. [CrossRef]
14. Giechaskiel, B. Effect of sampling conditions on the sub-23 nm nonvolatile particle emissions measurements of a moped. *Appl. Sci.* **2019**, *9*, 3112. [CrossRef]
15. Giechaskiel, B. Differences between tailpipe and dilution tunnel sub-23 nm non-volatile (solid) particle number measurements. *Aerosol Sci. Technol.* **2019**, 1–13.
16. Giechaskiel, B.; Vanhanen, J.; Väkevä, M.; Martini, G. Investigation of vehicle exhaust sub-23 nm particle emissions. *Aerosol Sci. Technol.* **2017**, *51*, 626–641. [CrossRef]
17. Giechaskiel, B.; Schwelberger, M.; Delacroix, C.; Marchetti, M.; Feijen, M.; Prieger, K.; Andersson, S.; Karlsson, H.L. Experimental assessment of solid particle number portable emissions measurement systems (PEMS) for heavy-duty vehicles applications. *J. Aerosol Sci.* **2018**, *123*, 161–170. [CrossRef]
18. Giechaskiel, B.; Lähde, T.; Drossinos, Y. Regulating particle number measurements from the tailpipe of light-duty vehicles: The next step? *Environ. Res.* **2019**, *172*, 1–9. [CrossRef] [PubMed]
19. Czerwinski, J.; Comte, P.; Mayer, A.; Reutimann, F. *Investigations of Changes of the 2-Stroke Scooters Nanoparticles in the Exhaust- and CVS-System*; SAE International: Warrendale, PA, USA, 2013; 12p.
20. Joint Committee for Guides in Metrology. *JCGM 200:2012 International Vocabulary of Metrology–Basic and General Concepts and Associated Terms (VIM)*, 3rd ed.; Joint Committee for Guides in Metrology: Paris, France, 2008.
21. Giechaskiel, B.; Clairotte, M.; Valverde-Morales, V.; Bonnel, P.; Kregar, Z.; Franco, V.; Dilara, P. Framework for the assessment of PEMS (portable emissions measurement systems) uncertainty. *Environ. Res.* **2018**, *166*, 251–260. [CrossRef] [PubMed]
22. Joint Committee for Guides in Metrology. *JCGM 100:2008 Evaluation of Measurement Data—Guide to the Expression of Uncertainty in Measurement (GUM)*; Joint Committee for Guides in Metrology: Paris, France, 2008.
23. Vasic, A.M.; Weilenmann, M. Comparison of real-world emissions from two-wheelers and passenger cars. *Environ. Sci. Technol.* **2006**, *40*, 149–154. [CrossRef]
24. Giechaskiel, B.; Cresnoverh, M.; Jörgl, H.; Bergmann, A. Calibration and accuracy of a particle number measurement system. *Meas. Sci. Technol.* **2010**, *21*, 045102. [CrossRef]
25. Takegawa, N.; Sakurai, H. Laboratory evaluation of a TSI condensation particle counter (model 3771) under airborne measurement conditions. *Aerosol Sci. Technol.* **2011**, *45*, 272–283. [CrossRef]
26. Giechaskiel, B.; Bergmann, A. Validation of 14 used, re-calibrated and new TSI 3790 condensation particle counters according to the UN-ECE Regulation 83. *J. Aerosol Sci.* **2011**, *42*, 195–203. [CrossRef]

27. Tsai, J.H.; Huang, P.H.; Chiang, H.L. Air pollutants and toxic emissions of various mileage motorcycles for ECE driving cycles. *Atmos. Environ.* **2017**, *153*, 126–134. [CrossRef]
28. Zardini, A.A.; Platt, S.M.; Clairotte, M.; El Haddad, I.; Temime-Roussel, B.; Marchand, N.; Ježek, I.; Drinovec, L.; Močnik, G.; Slowik, J.G.; et al. Effects of alkylate fuel on exhaust emissions and secondary aerosol formation of a 2-stroke and a 4-stroke scooter. *Atmos. Environ.* **2014**, *94*, 307–315. [CrossRef]
29. Zardini, A.A.; Suarez-Bertoa, R.; Dardiotis, C.; Astorga, C. Unregulated pollutants from tampered two-wheelers. *Transp. Res. Procedia* **2016**, *14*, 3109–3118. [CrossRef]
30. Clairotte, M.; Adam, T.W.; Chirico, R.; Giechaskiel, B.; Manfredi, U.; Elsasser, M.; Sklorz, M.; DeCarlo, P.F.; Heringa, M.F.; Zimmermann, R.; et al. Online characterization of regulated and unregulated gaseous and particulate exhaust emissions from two-stroke mopeds: A chemometric approach. *Anal. Chim. Acta* **2012**, *717*, 28–38. [CrossRef]
31. Spezzano, P.; Picini, P.; Cataldi, D.; Messale, F.; Manni, C. Particle- and gas-phase emissions of polycyclic aromatic hydrocarbons from two-stroke, 50-cm3 mopeds. *Atmos. Environ.* **2008**, *42*, 4332–4344. [CrossRef]
32. Prati, M.V.; Costagliola, M.A. Emissions of fine particles and organic compounds from mopeds. *Environ. Eng. Sci.* **2009**, *26*, 111–122. [CrossRef]
33. Zardini, A.; Clairotte, M.; Lanappe, G.; Giechaskiel, B.; Martini, G. *Preparatory Work for the Environmental Effect Study on the Euro 5 Step of L-Category Vehicles*; Publications Office: Luxembourg, 2016.
34. Varella, R.; Giechaskiel, B.; Sousa, L.; Duarte, G. Comparison of Portable Emissions Measurement Systems (PEMS) with Laboratory Grade Equipment. *Appl. Sci.* **2018**, *8*, 1633. [CrossRef]
35. Mohanta, L.; Iyer, S.; Mishra, P.; Klinikowski, D. *Uncertainties in Measurements of Emissions in Chassis Dynamometer Tests*; SAE International: Warrendale, PA, USA, 2014.
36. Geivanidis, S.; Samaras, Z. Development of a dynamic model for the reconstruction of tailpipe emissions from measurements on a constant volume sampling dilution system. *Meas. Sci. Technol.* **2008**, *19*, 015404. [CrossRef]
37. Mahadevan, V.; Iyer, S.; Klinikowski, D. Recovery of tail pipe species concentrations and its effect on emissions calculations from raw exhaust gas streams during chassis dynamometer tests. *SAE Int. J. Engines* **2016**, *9*, 1763–1774. [CrossRef]
38. Ajtay, D.; Weilenmann, M. Compensation of the exhaust gas transport dynamics for accurate instantaneous emission measurements. *Environ. Sci. Technol.* **2004**, *38*, 5141–5148. [CrossRef]
39. Weilenmann, M.; Soltic, P.; Ajtay, D. Describing and compensating gas transport dynamics for accurate instantaneous emission measurement. *Atmos. Environ.* **2003**, *37*, 5137–5145. [CrossRef]
40. Adachi, M.; Nakamura, H. *Engine Emissions Measurement Handbook: HORIBA Automotive Test Systems*; SAE International: Warrendale, PA, USA, 2014; ISBN 978-076-808-012-4.
41. Zardini, A.A.; Suarez-Bertoa, R.; Forni, F.; Montigny, F.; Otura-Garcia, M.; Carriero, M.; Astorga, C. Reducing the exhaust emissions of unregulated pollutants from small gasoline engines with alkylate fuel and low-ash lube oil. *Environ. Res.* **2019**, *170*, 203–214. [CrossRef] [PubMed]
42. Giechaskiel, B.; Zardini, A.A.; Clairotte, M. Exhaust gas condensation during engine cold start and application of the dry-wet correction factor. *Appl. Sci.* **2019**, *9*, 2263. [CrossRef]
43. Isella, L.; Giechaskiel, B.; Drossinos, Y. Diesel-exhaust aerosol dynamics from the tailpipe to the dilution tunnel. *J. Aerosol Sci.* **2008**, *39*, 737–758. [CrossRef]
44. Giechaskiel, B.; Arndt, M.; Schindler, W.; Bergmann, A.; Silvis, W.; Drossinos, Y. Sampling of non-volatile vehicle exhaust particles: A simplified guide. *SAE Int. J. Engines* **2012**, *5*, 379–399. [CrossRef]
45. Mathis, U.; Ristimäki, J.; Mohr, M.; Keskinen, J.; Ntziachristos, L.; Samaras, Z.; Mikkanen, P. Sampling conditions for the measurement of nucleation mode particles in the exhaust of a diesel vehicle. *Aerosol Sci. Technol.* **2004**, *38*, 1149–1160. [CrossRef]
46. Keskinen, J.; Rönkkö, T. Can real-world diesel exhaust particle size distribution be reproduced in the laboratory? A critical review. *J. Air Waste Manag. Assoc.* **2010**, *60*, 1245–1255. [CrossRef]
47. Giechaskiel, B.; Drossinos, Y. Theoretical investigation of volatile removal efficiency of particle number measurement systems. *SAE Int. J. Engines* **2010**, *3*, 1140–1151. [CrossRef]
48. Hawley, J.G.; Bannister, C.D.; Brace, C.J.; Cox, A.; Ketcher, D.; Stark, R. Vehicle modal emissions measurement—Techniques and issues. *Proc. Inst. Mech. Eng. Part D J. Automob. Eng.* **2004**, *218*, 859–873. [CrossRef]

49. Giechaskiel, B.; Lähde, T.; Schwelberger, M.; Kleinbach, T.; Roske, H.; Teti, E.; van den Bos, T.; Neils, P.; Delacroix, C.; Jakobsson, T.; et al. Particle number measurements directly from the tailpipe for type approval of heavy-duty engines. *Appl. Sci.* **2019**, *9*, 4418. [CrossRef]
50. Giechaskiel, B. Solid particle number emission factors of Euro VI heavy-duty vehicles on the road and in the laboratory. *IJERPH* **2018**, *15*, 304. [CrossRef] [PubMed]
51. Mamakos, A.; Schwelberger, M.; Fierz, M.; Giechaskiel, B. Effect of selective catalytic reduction on exhaust nonvolatile particle emissions of Euro VI heavy-duty compression ignition vehicles. *Aerosol Sci. Technol.* **2019**, *53*, 898–910. [CrossRef]

MDPI
St. Alban-Anlage 66
4052 Basel
Switzerland
Tel. +41 61 683 77 34
Fax +41 61 302 89 18
www.mdpi.com

Energies Editorial Office
E-mail: energies@mdpi.com
www.mdpi.com/journal/energies

www.ingramcontent.com/pod-product-compliance
Lightning Source LLC
Chambersburg PA
CBHW062105220926
51850CB00010B/2903
9783039286782